MW00425874

BERNARD ROMANS

Forgotten Patriot
of the American Revolution

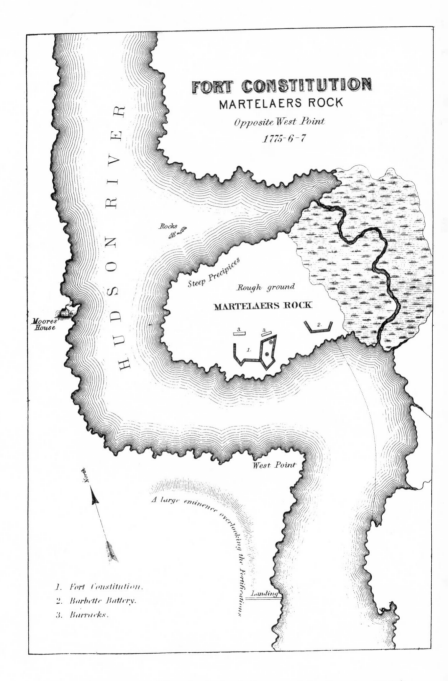

Fort Constitution: Martelaer's Rock (engraved for Boynton 1864).

BERNARD ROMANS

*Forgotten Patriot
of the American Revolution*

Military Engineer and Cartographer of
West Point and the Hudson Valley

By

LINCOLN DIAMANT

HARBOR HILL BOOKS
Harrison, New York
1985

Library of Congress Cataloging in Publication Data

Diamant, Lincoln.
  Bernard Romans : forgotten patriot of the American
Revolution.

  Bibliography: p.
  Includes index.
  1. Romans, Bernard, ca. 1720–ca. 1784. 2. Cartographers
—United States—Biography. 3. Military engineers—United
States—Biography. 4. United States—History—Revolution
—1775–1783—Cartography. 5. United States—History—
Revolution, 1775–1783—Engineering and construction.
I. Title.
GA407.R65D5 1985   526'.092'4   85–5421
ISBN 0-916346-56-0

Harbor Hill Books, P.O. Box 407, Harrison, N.Y. 10528

Typeset in 11 point Garamond by Type Express, Inc., Mamaroneck, N.Y.

For Rolf

Give the people their history.

Preface

History's pages are filled with great egos in search of satisfaction. Armed conflicts offer lurid opportunities; the American Revolution was no exception. In the violence of war, however, center stage provides little room. For each imperious general in the Continental Army (Charles Lee, Horatio Gates and Benedict Arnold, to name a few), hundreds of equally ambitious junior officers waited forever in the military wings.

Despite their desperate and often quite successful heroics, few of these volunteer midwives to the birth of America's New Age received even a footnote in history. But among those who did was colorful Captain Bernard Romans—although the superficial entries after his name in a few biographical reference books contain curious garblings.

My first acquaintance with this "ingenious man"* began while leafing through the thousand-odd pages of Volume 2 of the massive eight-volume series, *Naval Documents of the American Revolution.* There I discovered the 55-year-old (army) captain in the midst of a typical ruckus with his (civilian) superiors— apparently savoring every bit of his outrage. Romans projected an almost comic book impression; I soon found myself reading portions of the incident on the phone to my son in Boston, sharing mirth over Romans' misdirected indignation. It was not an auspicious beginning for what eventually turned into this small attempt, for sheer narrative pleasure, to retrieve part of America's past.

Governor Peter Chester of West Florida to the British Secretary of State for the American Department on August 14, 1772. [Additional footnotes do not occur in the remainder of this narrative; relevant sources are identified throughout the text, with further explanation in the Bibliographical Notes.]

The singular character of Bernard Romans, the belligerent participant in a two-centuries-old dispute, rose off the pages of the *Naval Documents* and led me forward. His inadequately told story was clearly an unusual and colorful thread in the otherwise familiar tapestry of early American Revolutionary history. As I scribbled and tapped away, I sought for some feeling of historical immediacy to better reflect the crisis time in which Bernard Romans pursued his fame and fortune. To that purpose, I decided to allow all who played a part in the Romans story to use their own words as much as possible, relying on direct quotation rather than paraphrase.

That decision reaffirmed my feeling that the pages of early American history are not filled by strange old people in three-cornered hats, but by recognizable personalities, driven by many of the ideas and ambitions that govern our own existence. In every age, heroism, devotion, sacrifice and herculean labor have always contended with arrogance, greed, selfishness and cowardice. In Bernard Romans' civilian career and his service among 200,000 other Revolutionary soldiers, one sees the same frustrations that can still arise from circumscribed abilities and unrewarded pretensions.

Research soon revealed Romans— who initially appeared nothing more than an egotistical blowhard quite out of his depth— to be an able if not fully-trained engineer, botanist, surveyor, seaman, explorer, linguist, cartographer, mathematician, writer, poet, artist and engraver— surely an American immigrant worth writing and reading about. In our own time, one commentator would even call him "a universal genius."

As this unusual man (of whom, peculiarly, no physical description or likeness exists) scrambled for positions of power in the pre-Revolutionary colonies and subsequent wartime chaos, he drove himself at a pace that usually outstripped both his talents and his purse— a risky combination. Struggling through a world of lesser men,

Romans, teetering between the heroic and the farcical, became a peripatetic thorn in the side of the Revolution. His difficult personality and "peculiar and bold opinions" usually left tension in his wake and generated a widely-held conviction that the engineer lacked the skills required for the posts he so eagerly sought. In retrospect, Romans' ill-starred contributions serve to contemporary advantage by throwing the earliest problems of our rebellion against the Mother Country into even higher relief.

The great tryworks of the American Revolution— like every major political upheaval— was not gentle with such a creative volunteer. It took this venturesome, querulous polymath— self-advertised in the *New York Mercury* in 1775 as "the most skillful Draughtsman in all America"; this early "war correspondent" whose patriotic output of up-to-the-minute money-making maps and engravings never faltered; and turned him, for a while, into some kind of military man.

Then it chewed him up, and spat him out.

Pondside
Ossining, NY

July 4, 1984

Table of Contents

Illustrations

Frontispiece: Fort Constitution: Martelaer's Rock (engraved for Boynton 1864).

13

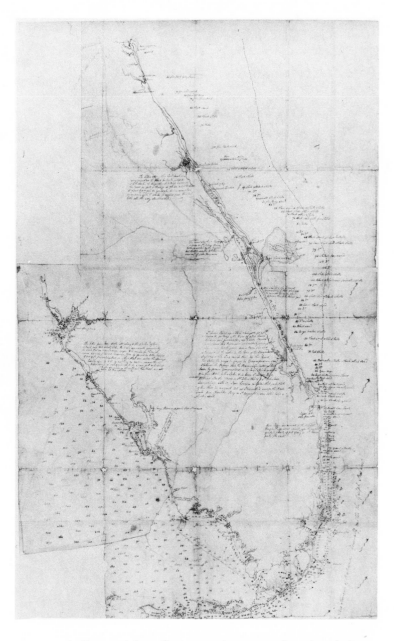

1. Romans' 6 sq. ft. manuscript map and mariners' chart
[1:750,000] of East Florida (c. 1770).

(William L. Clements Library, Univ. Mich.)

Introduction

In which Bernard Romans, energetic and productive deputy surveyor to the Southern District of British North America, falls into and out of favor with the Crown, and decides to seek new fame and fortune elsewhere in the colonies.

At best, Bernard Romans was a difficult man.

Born in Holland around 1720, he emigrated to England in his youth to study botany and mathematics, and soon graduated to the practice of engineering. In 1756, halfway through a rather undistinguished career, Romans decided to change his professional life. He accepted a civil service job as a junior surveyor in British North America.

Romans spent his next 17 years traveling widely by land and water throughout England's southern colonies as a "Draughtsm., Mathemn. & Navigatr.," also [he later claimed] making trips as far afield as Labrador and Central America. In this slow-to-develop region filled with tropical illnesses and various perils, Romans led a strenuous existence. As commander of various government survey vessels, he was not immune to maritime calamity. In 1766 he haplessly ran his fogbound hydrographic schooner "Mary" aground on "the south part of the Dry Tortugas" bank, and "in a second voyage lost her near Cape Florida, with about £500 ster. in her— a wound in my circumstances [he noted much later] as yet far from healed."

To Romans, those navigational disasters posed a challenge to his undisputed cartographic skill. From his own manuscripts and those of others, he began to compile an accurate set of marine charts and sailing instructions for all the "mazy" Florida waters— from the shallow Bahama Banks ("those Bug Bears to the fancy of our Navigators") as far

15

west as New Orleans. Through eventual publication of his
efforts, Romans hoped, "many a poor distressed Crew will
be saved from Ruin." He was right; in less than a decade,
his detailed materials would replace a century's accumula-
tion of inadequate and inaccurate aids to local navigation.

By the end of 1766, engineer-surveyor Romans finally
began to receive some of the recognition he so eagerly
sought. He was named one of several deputy surveyors of
Georgia, a post he held for the next two years. Inching his
way into the Southern establishment, Romans also tried his
hand at slavetrading and small-scale land speculation, but
with indifferent success.

In 1768, coincident with the Whitehall appointment of
Lord Hillsborough as both Secretary of State for the
American Department and president of the Board of Trade,
Romans was named principal deputy surveyor for the entire
"Southern District of British North America, and first
commander of the vessels on that service," at a salary of £30
a year. His district included the thinly-settled Floridas: at
the close of the French and Indian War, the Fontainebleau
treaty had "guaranteed in full right to his Brittanick Majesty
all that Spain possesses on the continent of North-America,
to the east, or to the south-east, of the river Mississippi."
(Later, under the treaty that ended the Revolutionary War,
the area would be restored to the patriots' Spanish ally for
an additional quarter-century.)

On October 7, 1763, George III created two new royal
colonies, East and West Florida (separated at the Apalachi-
cola River) from the former Spanish territory, bounded on
the north for almost 600 miles by the colony of Georgia.
But despite some unusual attention from London— all
government expenses borne by Britain and immigration
heartily encouraged— the seacoast's "fourteenth and
fifteenth colonies" were not flourishing. They had serious
geographic problems: heat, disease, poor quality land and
the inevitable hostilities with the natives. The information

that went forth about the Floridas to both North America and Europe was largely negative.

In 1768, Brigadier General Frederick Haldimand (later Governor General of Canada) was quoted in a letter by General Thomas Gage, politically influential senior military commander for the colonies, as insisting that "the Floridas will never repay half what is expended on them, either by the advantage of commerce or settlements." Gage said Haldimand assured him "there is nothing but sands and barrens from St. Augustine to Lake Ponchartrain, and no good land 'till you get near the banks of the Mississippi. The brigadier," wrote Gage, "seems convinced that there will be no settlers on the tracts which we now possess."

Moving into that vacuum, Romans became one of Florida's earliest promoters. Angry because active settlement of the area was taking place on less than ten per cent of the three million acres of munificent crown grants, he loudly criticized "the monopolisers of East Florida," who "overlooked the most useful Places there, and planted their Baronies in the pine barrens." Continuing on government service, Romans roamed the region's interior and sailed its coastline, working hard to alter the public image of the Floridas.

Painstakingly, Romans compiled detailed notes on the area's natural history, studying its natives and filling in empty places on existing maps and charts. "The Floridas," Romans wrote glowingly in one of his many reports to London, "certainly Bid fair to Become in a Small Space of Time as Flourishing a Country as any AMERICA can boast of." Never one to hide his own light under a bushel, Romans' observations were designed to call attention to the growing value of his own work.

As always, Romans was a stubborn man to cross. He soon was involved in a quarrel with his superior, John William Gerard de Brahm, a former military "Ingenier" under Emperor Charles VI. De Brahm was another strong-willed Dutch-English polymath who had emigrated to Georgia almost two decades earlier; in 1764 he became Surveyor General of the Southern District, headquartered in St. Augustine. And in 1770, finding himself "streightened in circumstances," he was unable to pay Romans' salary.

Romans overreacted by bringing an unsuccessful suit against de Brahm, and went so far as to accuse the surveyor general of "marks of insanity." As a result of that challenge, the cantankerous deputy was fired. He lost "above three-fourths of what I ought to have had," was indicted for poor performance by Georgia Governor James Grant (with whom he had previously crossed swords) and was subsequently discharged from government service by Lord Hillsborough in London.

Whereupon Romans decided he was "absolutely at liberty to make use of observations" he had acquired through any of his official duties. [His decision was the same as Thoreau's 70 years later: "I will still make what use of the State I can, as is usual in such cases," but it would publicly return to plague the engineer.]

Removed from the Whitehall payroll, Romans fell back on what he did best. "Entirely at my own vast Expence and bodily Fatigue," he relates, he embarked on several "long and tedious" expeditions of chartmaking and botanizing. He sailed from St. Augustine through the Bahama Islands, and then ran along the coast of West Florida as far west as the latter's boundary (at that time) on the Mississippi delta.

Returning to Pensacola in August 1771, Romans—always resourceful— was able to find new local employment. He was assigned by the influential and wealthy John Stuart, Superintendent of Indian Affairs for the Southern District, to survey the West Florida interior. Although it had been previously explored and was capable of producing many valuable crops for the mother country, most of this criticized but rich Florida area was still unmapped.

First charting the bays of Pensacola and Mobile, Romans and his companions then set off by canoe for a journey through the rainy winter of 1771-1772, penetrating the Choctaw and Chickasaw nations to the northwest [now part of Alabama, Mississippi and Louisiana]. The party crossed into west Georgia and worked up the Tombigbee River to its headwaters near lat. 33° 30'. Besides a detailed and interesting frontier journal, Romans' four-month expedition produced a "tolerably accurate map based largely on Indian hearsay" (according to an early 20th century historian), which suffered by "blending two separate streams into one, and failing to lay down creeks of the largest size."

It contained (the critic also noted) "many clerical slips in the names" of the 60 Choctaw towns Romans had indicated— "all the evidence shows that Romans had but little knowledge of the Choctaw language, and wrote many names as they sounded to his ears." For example, *kushak osapa* is rendered as *Consha Consapa,* and "the farther the map gets from Romans actual line of travel, the more confusion."

But none of this was either apparent or very meaningful at the time; Romans was officially hailed for his ability to erase empty spots on the map, removing those "uninhabitable downs, with elephants for want of towns." On his return to Pensacola in January 1772, West Florida's Governor Peter Chester gave him an additional job mapping new areas to the northeast. Romans earned special praise from Chester by completing that particular draft of "parts not hitherto explored" in less than three months.

On August 14th, Governor Chester forwarded the new map to Hillsborough, noting that it had been "executed in consequence of my direction by Mr Bernardus Romans, a surveyor who I think very capable of performing business of this kind. I also transmit some of Mr Romans' draughts of flowers etc. and a specimen of the jalup which he has lately discovered within the province."

The somewhat impoverished surveyor was surely dazzled by Governor Chester's conclusion: "As this Mr Romans appears to be an ingenious man and both a naturalist and botanist, I think him worthy of some encouragement, and

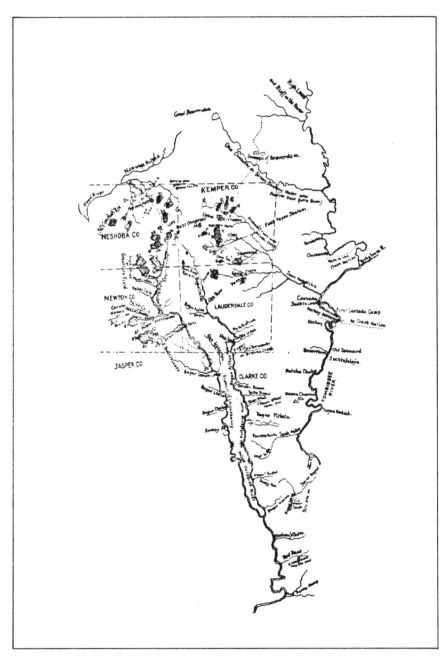

2. Romans' 1772 *Tombigbee River Basin* map (from a 19th century copy adding Mississippi county lines).

submit to your lordship whether it would be improper to give him a salary of £50 or £60 per annum, added to the estimate [West Florida's annual budget] or otherwise, in order to induce him to continue in this colony to make discoveries and observations in botany, some of which may probably hereafter be of use to the public."

What Hillsborough might have thought about rehiring the man he had fired two years earlier is immaterial. The Secretary of State had himself been replaced at Whitehall the day before Governor Chester sent off his letter.

In the Floridas, Romans continued busy. Working from his own notes and those of two other capable surveyors, George Gauld and Lieut. Governor Elias Durnford, Romans reworked their mutual explorations into a giant 40-square-foot *General Map of West Florida— the Whole Examined and Carefully Connected at Pensacola on August 31, 1772.* Governor Chester, passing Romans' work on to London via New York and General Gage, praised it as "a more perfect map of the province than any hitherto transmitted." (When finally deposited in the British Colonial Office the following July, it had acquired the cryptic endorsement: *Q. If this is not Mr. Romans' performance.*)

On this map and in the important description that accompanied it, Romans recorded coastal soundings; located rivers, settlements and native villages; indicated soil types and crops; described local customs and practices; and named the area's fish, minerals and medicinal plants. In short, Romans attempted to set down every bit of information that might prove useful to the British Government in further "opening up" the country. As a calculated effort to refurbish his official ties, it was wholly successful.

A born-again Pensacola booster, Romans viewed with disdain the extensive Spanish fisheries in West Florida waters— "three or four Hundred White Men" shipping more than 1,000 tons of dried salted fish 90 miles to Cuba between September and March of each year, he said, "while we of both Floridas See and Sit Still." Romans urged complete expulsion of the Spanish, in favor of local English fishermen— "not a Visionary Scheme, as Governor Grant once was pleased to say when I related all this."

If it came to frightening Spain, Romans was ready with tougher ideas. "How noble is the sight of this Coast," he wrote, "for the rendezvous of a fleet of Men-of-war to command the Spanish money. Five Harbours that are able to receive as large Ships as are necessary in this part of the world, make a Chain from the southeast to northwest and west. Winds and Currents, through the wonderful province of the Almighty, will not permit the ships with the Spanish treasure to come any other course or way than very near this chain of Harbours, so that by intercepting these, the power of Spain may be guided (nay commanded) by Great Britain.

"The Treasure that comes this way," Romans noted, "is lately increased in a prodigious proportion; for since the laying aside of the port of Acupulco by the Spaniards, all the money which used to go that way to Manilla now finds its way to Europe through this Channel. But," he concluded with proper modesty for someone urging piracy or war on a country that had been at peace with Great Britain for almost a decade, "I confess myself not politician enough to say more on this head, and I therefore leave it to abler judges."

For the remainder of his report, Romans was particularly scornful of the character of the Florida natives, and moreso of the whites who came overland from Carolina and Georgia to trade with them. "A great deal has been said by Many Writers," Romans advised the Colonial Office, "that the natives are a very good, Simple, honest and even virtuous people. I must Contradict them, and Say that they are Treacherous, Cruel, deceitfull, Faithless and thieving."

As for the traders themselves, Romans wrote evenhandedly, "the Behaviour of the Vile Race who now Carry on that Business, is such that a relation of it in the most favourable Manner could not fail to Shock humanity. Nay, the very Savages are scandalized at the Lives of those Brutes in human shape, the very scum and outcast of the Earth, always more prone to savage barbarity than the savages themselves." Romans was hardly a man of easy opinion.

In August, Romans also began a lopsided correspondence with John Ellis, for eight years London's royal agent for the Province of West Florida. Romans suggested that Ellis underwrite the establishment of a royal botanic garden in Pensacola that could contain "every curious and usefull plant that Grows from the Capes of Florida to Canada[!], besides Jamaica and Cuba."

In addition to his government responsibilities, Ellis was also a famous botanist and founder of the British Linnaean Society. He enjoyed the confidence of amateur botanists throughout British North America, and regularly received from this new land many "curious subjects in Natural History." Thus Romans' letter was not unusual, but Ellis was initially uneasy about this Florida stranger— who also ingenuously enclosed his personal "Observations on a Catalogue of Plants Published by John Ellis Esqre. F.R.S." With his official cartographic work in the Floridas now completed, Romans required a fresh source of income. The suggestion of a royal botanic garden to John Ellis was merely one possibility.

In December 1772 Ellis asked his Charleston botanist acquaintance Dr. Alexander Garden (eponym of the gardenia) for additional information about Romans. Meanwhile fortune finally rewarded the ambitious engineer; with Hillsborough dropped from favor, the new British Secretary of State Lord Dartmouth wrote to Governor Chester from Whitehall on December 9th that he would have "no objection to provision for Mr Romans on the estimate." It took almost four months, however, before the Plantations Office moved to approve the lower of Governor Chester's two suggested salaries. When Undersecretary of State William Knox formally issued the 1772-1773 West Florida Estimate on March 3rd, 1773, £50 of its £7274 13s. 6d. total was "allowed to Mr Romans for care and skill in the collection of rare and useful productions in Physick and Botany."

However, Romans was denied any knowledge of this delayed small beneficence, which henceforth would help to keep wolves from the door, until he had passed through a new employment crisis. In February 1773, Romans recounts how he was "deluded" by his former patron John Stuart into coming to Charleston, fourth largest city in the colonies, to accept a new £150-a-year post as the "King's botanist." The income promised positive security. But en route, Romans' ship *Liberty* "overset" in very rough weather off present-day Palm Beach; he lost his extensive botanical collection, with the exception of "a very few seeds which I later gave to Dr. Gardner [Alexander Garden]." The delay caused by the accident proved fatal to the new proposal; when Romans finally arrived at Charleston, Stuart's lucrative job tender was withdrawn.

To Romans, angry and confused, it was clearly the time for a new direction. For almost two decades, the engineer had roamed and mapped the American Far South, assiduously accumulating useful and valuable information on all of its geography and natural history. By now this venturesome but impecunious man— presumably unmarried—was 53 years old. He was not exactly well; for almost a decade he had been swallowing opium— on occasion, he tells us, "in incredible doses"— for the "haemorrhoidal flux" (an obstinate tropical dysentery).

Yet despite all this Romans still possessed seemingly boundless practical energy; and was even dreaming of a four-year trip by boat and on foot around the world. He responded swiftly to Stuart's change of mind by taking passage north with his slave for what he hoped would be more appreciative latitudes.

Sailing on the same Charleston tide, a London-bound vessel finally carried to John Ellis a negative appraisal of Romans from Alexander Garden: "A man but little versed in Botany; he knows a few of the terms, but his knowledge in the science, as far as I can judge, is both much limited and very superficial. He has been employed by the superintendant of Indian affairs as a surveyor, and I have seen a few things of his collection, but they were of no account or value. He is at present out of employment, and is here on his way to New York, where he intends to publish a chart of the Bahama Islands and Cape Florida.

"He appears to me," concluded the eminent Dr. Garden (who would flee to England during the Revolution), "to be a man who would take pains, but I am afraid he wants knowledge." Unsaid was Garden's customary 18th century view that science best consists of clearly separated disciplines. Romans' attempts to make himself important in too many fields at once must have made the wealthy doctor uneasy.

Unspoken was the fact that Garden, De Brahm and Stuart were all good friends; seven years earlier they had joined a fort-building expedition into Cherokee country across the Great Smoky Mountains. Undoubtedly they shared an establishment prejudice against parvenus. That identical feeling was mirrored by a Charleston newspaper— such self-made professionals, it said,were engaged in "one continual Race in which everyone is endeavoring to distance all behind him; and to overtake, or pass by, all before him; everyone is flying from his inferiors in Pursuit of his Superiors; who fly from him, with equal alacrity."

Bernard Romans, now en route to New York, was able to put all of this eight hundred miles behind him. His world seemed brighter. But at that same moment, King George III in London was about to offer consent to an apparently innocuous Act of Parliament. For less than £300 in annual revenue, the British government was giving the East India Company the right to monopolize the tea trade to the colonies. Eventually, the measure would become an excuse to turn every independent American's life— including Bernard Romans'—inside out.

Thursday the 22 July 1773

Ordered that a Certificate be Given to B. Romans Sign'd by this Committee in Order to give Sanction to his draughts, as far as the Committee Appointed have gone through in the Examination of them. ——

Voted that the sum of Fifty Pounds be paid to B. Romans to Enable him to Execute his Design, for which sum of Fifty pounds he is to give Bond to the Society for to lodge with the Society a Copy of his draughts with a Book of Directions for the same. ——

At a meeting of the Committee of the Marine Society on Monday the 2d August 1773 at the House of Thomas Doran

Pres.t

Samuel Tudor	Parch.l N. Smith	Absentees
Vin.t R. Stilfield	Patrick Dennis	Sam.l Bayard – Excus.
Jon.a Lawrence	Isaac Sears	Ja.s Creighton – d.o
Alex.r M.c Dougall	Jacob Roome	
And.w Griffiths	Benj.a Davis	
Nich.s Fletcher		

The following Persons were proposed members of this Society
Cap.t Bernard Romans – Cap.t Eben.r Turell — Cap.t Edw.d Hopper
Cap.t Joseph Stringham — Cap.t Tho.s Richardson — Cap.t W.m Cannon
John Wood Esq — — — Chris.r Duyckinck
John Dawar — — — George Bell —
Who being severally ballotted for were declared Members

3. 1773 minutes of the New York Marine Society: July 22 — advancing Bernard Romans £50; August 2 — electing him a member.

CHAPTER I

New York & New England

*In which Bernard Romans joins two influential
societies, seeks royal sponsorship for a proposed
trip circling the globe, replies bitterly to
accusations of plagarism, publishes his book (and
enormous map) on Florida, and sees the Revolution begin.*

Bernard Romans sailed into the port of New York from
South Carolina in the early summer of 1773. Ready for a
fresh start, he made immediate contact with the press,
ambitiously announcing he sought "the Assistance of some
Gentlemen (Lovers and Encouragers of American Literature)
towards Publication" of a long-dreamed-of Florida carto-
graphic/literary project.

Romans was planning to engrave a new pair of large maps
of that entire area to accompany a *Concise Natural History of
East and West Florida.* Joint publication of the book and
maps, Romans' newspaper ads declared, would be "as
desirable for the Sage in his Cabinet as for the Mariner in his
Ship." [Part of his work did indeed become a basic
mariners' guide to the "Bank of Bahama, the coast of the
two Floridas, the north of Cuba and the dangerous Gulph
Passage." That very practical section, also republished
separately in 1789, went through four London editions in as
many years.]

To interested readers in the northern colonies, Florida
was still a land of primitive mystery. Despite the growing
economic uneasiness on the eve of the War of Independence,
Romans was able to obtain a respectable number of
subscribers in advance of publication, including New
York's influential Provincial Governor William Tryon.

More than a quarter of his subscribers were sea captains,
suggesting that Romans' huge small-scale charts, carrying

vessels from coastal sounding to coastal sounding across
600 miles of latitude and 900 miles of longitude, were
eagerly sought after. Even the office of the Royal Engineers
in New York City ordered six copies. With some advance
money in pocket, Romans' tedious job of typesetting his
manuscript for the initial 440-page volume, "Illustrated
with twelve COPPER PLATES, And Two whole sheet MAPS, "
could begin. Eight of its plates would be Romans' own
crude attempts at engraving Florida plants and natives.

This unusual engineer from the South— self-proclaimed
as a correspondent of the famous botanist John Ellis—
found himself welcome in intellectual circles of the bustling
New York metropolis of more than 20,000 people. Within
a month of his arrival, Romans had solicited the influential
Marine Society— a charitable and educational group of
local sea captains [today in its third century]— for financial
assistance to speed the engraving of the expensive whole
sheet maps that would accompany his *Concise History.*

A committee of the Marine Society, including Captain
Samuel Bayard (whose path would cross turbulently with
Romans' only two years later) inspected his drafts. They
recommended Romans be lent £50 without interest, under
bond of a set of copies of his Florida manuscript charts and
sailing directions (*Figure 1* probably represents part of that
"security.") In August the Society admitted him as Member
No. 485, and his fellow members suggested he contact the
talented Boston engraver Paul Revere.

Romans was drafting his pair of huge maps (27½ and 13½
square feet, respectively) to straddle the Florida peninsula
and panhandle (then reaching northwest to the Mississippi)
like a fat, inverted "L." Only one engraved copy of each
chart still exists, in the Library of Congress. Time does not
always look kindly on huge old maps. Purely decorative
items may last forever, but accurate sailing charts soon wear
out with use. The upper whole sheet map was "Humbly
inscribed" to his "kindest patrons," the New York Marine
Society.

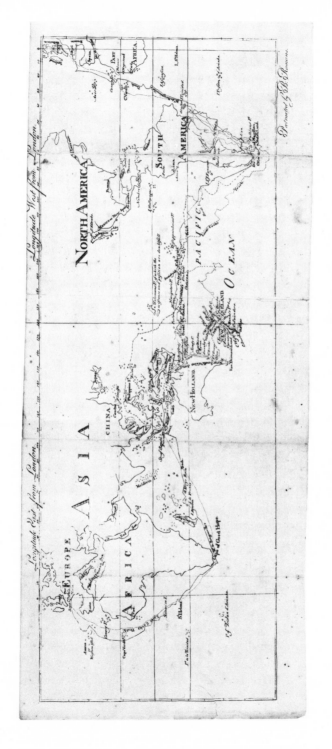

4. Romans' engraving of *Lieut. Cook's tracks* for publisher James Rivington's edition of Hawkesworth's *New Voyage* 1774.

For additional income, Romans commenced an ongoing relationship with James Rivington, one of New York's busiest printers, "protracting" a *Map of the World Showing Longitude East and West from London* for Volume I of Rivington's American copy of John Hawkesworth's *A New Voyage Round the World in the years 1768, 1769, 1770 and 1771. Performed by James Cooke* [sic] *in the ship Endeavour.*"

This reprint of the 1773 London edition that described Cook's epochal scientific circumnavigations included additional engravings by the ubiquitous Paul Revere, and was one of the most widely advertised books in the colonies; it carried a 17-page list of subscribers, including Romans himself. The engineer was also pleased to see his credit, "Botanist to his Majesty in West-Florida."

Invited to Philadelphia in August, Romans addressed the American Philosophical Society (founded four years earlier by Benjamin Franklin). He had corresponded with the Society from Pensacola, and to them he now described in person two Florida plants (dismissed by one member as "nondescript"), and read a paper on recent Dutch improvements to the mariner's compass. He displayed a finished draft of his lower whole sheet map, showing East Florida, part of the Bahamas and the northern coast of Cuba. That sheet contained a flowery dedication, within a pair of elaborate if awkwardly-drawn cartouches, to "The HONble. the PLANTERS in Jamaica and all Marchants concerned in the Trade of that Island, being chiefly interested in the Navigation herein explained," as well as "All COMMANDERS of Vessels round the Globe."

Although the earliest Spanish and French maps of the area dated back to 1527 (with a particularly handsome version by Theodor de Bry 175 years prior to Romans), Romans was the first cartographer to compile a chart of the entire Florida coastline based on meticulous personal observation and exploration. His two maps included more than a thousand coastal soundings, "more particularly laid down," Romans boasted, "than any Map Compiler has ever

5. Cartouche detail (reduced ⅓) from Sheet II, Romans' *Part of the Province of East Florida* 1774, engraved by Paul Revere.

Dr. James Lloyd, of Boſton.

M

Mr. Archibald M'Clean, of York county.
Capt. John Matreſon, of New-York.
Timothy Matlack, Eſq.
Dr. George Millegan, of South-Carolina.
James M'Clurg, M. D. Williamſburgh, Virginia.
Mr. Benjamin Morgan.
Rev. Dr. Samuel Magaw, Vice-Provoſt of Univer. Phila.
Hon. Dr. James M'Henry, Eſq. of Baltimore.
Rev. Dr. James Madiſon, Preſident of the College of William and Mary, Virginia.
Rev. Mr. Henry Muhlenberg, of Lancaſter.

P

Dr. John Perkins, of Boſton, Maſſachuſetts.
Dr. Thomas Park.
Mr. Robert Patterſon, Prof. Math. Univer. Philadelphia.
Hon. Mann Page, Eſq. of Frederickſburgh, Virginia.
Thomas Paine, Eſq. of Bordentown.
Charles Pettit, Eſq.

R

Mr. Bernard Romans, of Penſacola.

S

Dr. Hugh Shiell.
Jonathan Bayard Smith, Eſq.
Jonathan Dickinſon Sergeant, Eſq.
Rev. Dr. Samuel Smith, Vice-Preſident of the College in New-Jerſey.

T

Dr. James Tilton, of Dover.
Mr. John Ternant.

V

Samuel Vaughan, Eſq.
Mr. John Vaughan.

W

His Excellency General Waſhington, Virginia.
Rev. Samuel Williams, L. L. D. Hol. Prof. Mor. and Nat. Philoſ. College of Cambridge, Maſſachuſetts.
Dr. Nicholas Way, of Wilmington.
George Wall, jun. Eſq. of Bucks county.
Hon. Anthony Wayne, Eſq. Gen. in the Armies of the United States.
Mr. Benjamin Workman, Teacher of Math. Univerſity, Philadelphia.

6. Partial list of new American Philosophical Society members from 1771 to 1786 (Paine, Romans, Washington, Wayne, etc.).

had an Opportunity to do." The Philadelphia Society was as
impressed as we still are today. Dr. Hugh Williamson, a
learned member of the Society's Council who had taken his
medical degree in Holland, sponsored Romans' application
for membership. Five months later he was elected— at the
same time as John Ellis— to the most influential scientific
body in America.

On his return to New York, Romans— ever ready to
undertake many ambitious projects at once, and now
seriously "streightened for want of cash"— dispatched an
ingratiating (but prophetic) letter to the Secretary of State
for the Colonies, William Legge, 2nd Earl of Dartmouth,
and half-brother of Britain's irresolute Prime Minister Lord
North. Lord Dartmouth, termed "a truly good man" by
Benjamin Franklin, had demonstrated a genuine interest in
colonial progress. Before he wrote, Romans had learned
with delight from friends in Florida of Dartmouth's
approval of his £50 Florida salary; but possibly embarrassed
by the fact that he was no longer "continuing in the
colony," he immediately began to call that payment a
"pension," and took an unusual liberty with John Ellis by
drawing his first money against Ellis' personal credit in New
York.

Romans' letter to Dartmouth— whose official papers
would later yield up a map of the historic boundaries of
New York Colony (signed "B. Romans," but displaying
little of the engineer's calligraphic style)— offered a striking
combination of 18th-century grantsmanship, coupled with
an unusually candid personal analysis. Romans began with
the traditional salutation to the royal establishment—
obsequious thanks for having been "drawn out of the
Obscurity in which I was hid," plus a proper expression of
humility and "a due sense of my slender abilities, yet at the
same time a Consciousness of my readiness to Exert them
to the utmost." It was vintage Romans, and accurate. Then,
inspired by his engraving of Cook's circumnavigation, the
King's surveyor made a dramatic proposal:

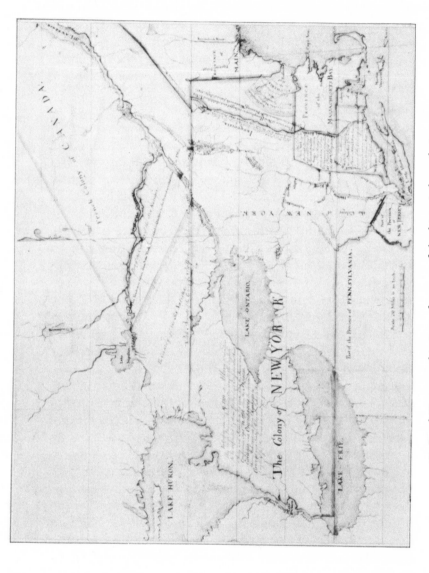

7. Lord Dartmouth's 3 sq. ft. map of the historic boundaries of New York Colony [c. 1773, attributed to "B. Romans"]

"The general knowledge I have obtained of the Geography of this Extensive Continent in my repeated travels through its vast wilds," he wrote fearlessly, "have made me form an Idea, that a Journey into the North Eastern parts of Asia would not be attended with the numberless difficultys hitherto objected. With the least Assistance to the Expence of the Journey, I would undertake it. May I then presume to intreat your Lordship to lay this my humble proposal before our most Gracious Sovereign [George III].

"Upon a favorable reception," Romans continued, "I will instantly remit a Plan of the scheme, and with impatience wait for the welcome orders to proceed on this desirable Discovery. I do believe that no man could do it on a more frugal plan ['frugal' was Romans' favorite adjective whenever discussing public expense]." The first stage of his projected journey, boldly crossing the North America continent on foot (shortly before Captain Cook's continuing explorations carried him along the Pacific Northwest littoral) foreshadowed by three decades Lewis and Clark's own historic 1804-1806 traverse.

But Romans was dreaming of an even grander conclusion to his proposed circumnavigation, as he packed his saddlebags to ride north to Boston. That noisy little seaport's 17,000 inhabitants were still seething from the previous month's Tea Party— fallout from Parliament's ill-timed legislation. In Boston Romans was planning to call on engraver Revere, and also promote his book by "putting up & distributing printed Proposals through the Town," with ads in the *Gazette and Country Journal* seeking additional subscribers among Boston's "Well-wishers of American Science and Navigation." It was a busy schedule.

The first sheets of the *Concise Natural History,* Romans told Boston newspaper readers, were coming off press in New York, printed "on a very good paper." He wished, he said, to "return to compleating the Work with redoubled Ardour; the ordinary thought of its falling through will not so much as be suspected."

Soaring beyond these traditional concerns of book promotion, Romans was also dreaming of ways to project his proposed trip far beyond "the North Eastern parts of Asia"— all the way around the world to London! From Boston on December 20th he wrote to Dr. Williamson at the Philosophical Society on "the route to be taken, either by the Lakes or by the Mississippi to the Siaoux; by the mississippi preferably, being less known. Mr. Adams & my son who is thoroughly acquainted with the Woods and 12 other men to accompany me." [With that sentence, Romans gives a surprising indication of fatherhood, but never refers to a grown son again.]

"One of these 12," Romans continued to Williamson, "should be a person of some Education and capable of reconducting 4 of them after our arrival on the Western Shore of America, carrying with him a Copy of our Journals. My self with the remaining Nine to proceed Northward untill by some means or other we may arrive in the North Eastern parts of Asia, & this through Siberia into Russia, thence through Poland, Germany, Holland & France, to make my report in Great Britain.

"The first summer," Romans went bravely on, "will be spent in reaching the Siaoux, where we must Winter the first season. The next Summer we may be morally Certain of finding a Water Course running West Ward which will lead us to the Western Shore of America, where we must pass the Winter of the second year. The third spring the above 5 men must be Sent Back with a report of my proceedings to Philadelphia, & my self continue my route North Westward. I make no doubt that this summer would suffice to carry me to the North Eastern parts of the russian Empire, where we would stay the third winter.

"The fourth summer I guess would carry me to Petersburg. I must be furnished with a Russian pass, & Letters of Credit for the different States I am like to pass through." Romans also proposed trinkets for natives, boat-building tools—and high wages for all on his expedition, with money left behind for family support. All things considered, it was a grandiose scheme at a highly inopportune time; the colonies' intemperate reaction to new ministerial taxation was serving to focus George III's attentions on far more compelling problems than a heroically-led expedition around the globe.

Amid Boston's growing political turmoil— despite his brand-new Crown salary and eagerly-sought-after royal support— Romans began to revise his manuscript of the *Concise Natural History* to reflect a more aggressive patriotic flavor. Reflecting rising tensions in the Massachusetts capital, Dr. Joseph Warren was already calling for "resistance to the unparalleled usurpation of unconstitutional power, disarming the parricide which points the dagger to our bosoms." Romans enlarged his manuscript comments on the practical potential of growing tea in Florida, to recharacterize that plant as "a despicable weed, and of late attempted to be made a dirty conduit, to lead a stream of oppressions into these happy regions."

Romans commissioned Paul Revere to engrave the two huge whole sheet maps of East and West Florida that would accompany his little octavo volume. "The elegance of the Map, added to its large Size," Romans advertised, "will render it an ornamental Piece of Furniture." Later that summer Revere billed him £17 for his work in helping create the maps.

Romans sold two full subscriptions to a man who would one day become a very influential friend, John Hancock (who earmarked one for Harvard College). He took an order for 50 books without maps from general merchant and proprietor of Boston's "London Bookstore" Henry Knox, and set aside some extra copies for Hybertus Romans, a relative in Amsterdam, and for John Ellis. He also took advantage of his stay in Boston to extract two sections from the *Concise Natural History* dealing with Florida cultivation of indigo and madder, and arranged for publication of both in the new *Royal American Magazine,*

published by Revere's close friend the well-known peripatetic printer Isaiah Thomas.

Thomas' January 1774 issue also printed this poetic
contribution from Romans on "The Worth of AMERICA."
Reflecting an elegaic style that seems somewhat out of
keeping for the engineer, its 120 lines include:

> *No land gives more employment to the loom,*
> *Or kindlier feeds the indigent; no land*
> *With more variety of wealth rewards*
> *The hand of labour: thither from the wrongs*
> *Of lawless rule, the free-born spirit flies;*
> *Thither affliction, thither poverty,*
> *And arts and sciences; thrice happy clime*
> *Which Britain makes th' asylum of mankind.*

Off to Newport RI to seek additional subscriptions,
Romans impressed local pastor Ezra Stiles with his extensive
knowledge of the Southern native tribes of "Cherokees,
Chauktaws, Creeks, Chickesaws and Catawbas." Stiles
wrote approvingly in his diary: "Captain Romans has
travelled among all the Indians from Labradore to Panama
[Romans had also told Ellis he had been to Colombia]."
Romans described Labrador in somewhat strange terms to
Stiles— "the Esqimaux have beards & these pretty large."
The president-to-be of Yale College subscribed to neither
book nor map, although Romans' advertising now confi-
dently warned that "none will be had for less than 16
Dollars after publication."

Returning through Connecticut, the colony in which he
was soon to settle, Romans arranged with gifted 33-year-old
New Haven silversmith Abel Buell— America's first
typefounder and a reformed "C"-branded and ear-cropped
counterfeiter— to engrave three small maps of Mobile and
Tampa Bars and Pensacola for the appendix of the *Concise
Natural History.* By February 1774 Romans was back in New
York, in time to angrily refute allegations published (and
what was worse, "mentioned to the Marine Society") in his

absence by one John Scott, "who had lived under the same roof" with Romans in Florida. Scott claimed the author/ cartographer of the soon-to-be-published *Concise Natural History* was pirating the work of other writers and mapmakers.

Vehemently attacking "busy body" Scott in two long advertisements in Rivington's *New-York Gazetteer,* Romans declared him "a SCOUNDREL, who knew better than to mention this vile inuendo ... but defamation always finds some rascally conduit, through which, like a vile cur, to bark at innocence." If he ever needed help with his mapping, Romans insisted (presumably with Gauld, Durnford and others in mind) that he had "made honest applications for assistance. I have not pilfered nor pirated from any man living."

As to his literary efforts, Romans took "appeal to Mr. Rivington and Mr. Hazard [Ebenezer Hazard, his distinguished fellow member of the American Philosophical Society], whose acquaintance with authors entitle them to be competent judges of books, whether they know of anyone from whom I have printed one paragraph." To the now typeset appendix of the *Concise Natural History,* Romans added this bitter observation: "I have done more than was desired of me; which, however, has never been the occasion of any favor being bestowed, any more than 'you did well, and I am glad you did it.' Therefore I am under no manner of obligation to any of the LITTLE GREAT ONES who have occasionally used me (sometimes as the monkey did the cat)." It was a characteristic display of anger at those in power for his unreciprocated attentions.

Feeling he had placed all "stain or imputation" loudly behind him, Romans turned his attention to practical problems of getting the *Concise Natural History* through the press. He tried to move faster. It was the spring of Parliament's "Intolerable Acts"; the country's interest was being forcibly shifted towards far graver matters than the engineer's opinions on the future potential of the Florida frontier. On February 28th, Romans wrote to William

Bradford, the well-known Philadelphia printer (and sub-scriber to the *Concise Natural History*) to ask "whether paper fit to print the maps on may be had there, and on what terms?" Romans' own success in selling 75 copies to bookstores in Boston enboldened him to urge Bradford to "ask some of your Philadelphia booksellers to follow"; there were 50 active bookstores in that city.

Demonstrating an obvious fear of the economic inse-curities of authorship, Romans also struck out wildly with new letters to John Ellis, who had long since ignored Romans' suggestion to fund a Pensacola botanic garden at $241 Spanish dollars a year, with salary and traveling expenses for the director carefully left unstated. In March 1774 Romans complained to the King's Florida agent about Dr. Alexander Garden's past aloofness— and then again, in May, wrote that "the present troubles in America Leave me little hope of my proposal to Lord Dartmouth being taken notice of, concerning a Journey through America to Asia. This grieves me much, as I live in a part of the World where the Study of Nature and its votaries is in a most unaccountable manner Neglected, & I have but Little Else to recommend me to the attention of mankind.

"I Lead a very neglected Life and am very hard put to it to maintain myself," Romans complained to the man to whose account he had charged, without authorization, his initial salary payment. "As I have no friend in Europe to whom to apply," he wrote, "I once more take the freedom to address you on that head, hoping it may be your inclination to recommend me to Some place or business, Be it never So trifling. I will Strive to show my gratitude."

Romans ended with a plea for Ellis' support in obtaining at least a royal monopoly for curing and exporting the jalup (*jalap*)mentioned by Governor Chester, a purgative plant at one time thought to be exclusive to Mexico, but which Romans had discovered in 1772 growing not far from Pensacola, "the first finding in any settlement Belonging to Great Britain."

Unbeknownst to Romans was the happy fact that Whitehall had just approved continuing his £50 salary as part of Governor Chester's 1773-1774 £4850 West Florida Estimate— despite the engineer's removal to New York. Meanwhile, he set to work to engrave the handsome calligraphic copper plate that would dedicate his entire *Concise Natural History* to John Ellis. Never one to leave only a single iron in the fire, Romans also began to gather original materials for a new book on the Netherlands, a two-volume political and military history of the country of his birth, since the 1500's. And soon he was able to announce in the *New-York Gazetteer* that plates for his large whole sheet maps to accompany the *Concise Natural History* were finally engraved. Printing awaited only the arrival of a specially-made paper, he said; it had been ordered on Bradford's recommendation from the respected Wilcox mill near Philadelphia.

By the spring of 1775, the book itself (but not the accompanying maps which, despite the £50 advance from the Marine Society, remained unprinted for many months) was finally bound and ready. Romans attributed one last delay to problems with the illustrations— a "struggle of 4 months occasioned by the Art and Mystery of Copper-plate printing"— plus this additional apology: "*The Map of the Savage nations,* intended to be put facing page 72, was engraved by a gentleman [Abel Buell] who resides in the country 60 or 70 miles from New York, to whom the Map was sent, but it was sent back and miscarried through the carelessness of the Waggoner. Though the publication has been delayed some time on that account, the Plate is not yet to hand. The reader will therefore please to expect said Map with the second volume."

Rivington's newspaper carried a final plea from Romans: "The expence of this Work having much exceeded expectations, it is very necessary after so great a drain of money, that the subscriber immediately send orders on where it is to be delivered, with no demur being made in payment so the author may at last receive the return of his labor." Copies could go forth to subscribers— and also to New York's 16 booksellers; Bernard Romans undoubtedly made certain they did.

Even minus its miscarried plate and long-promised whole-sheet maps, the *Concise Natural History of East and West Florida* was a success. For a writer who still claimed uneasiness in his adopted language— "no elegance of style, nor flowers of rhetoric, must be expected"— the wide variety of natural, aboriginal and historical information and speculation that Romans was able to cram into his fascinating little book marked it an important literary work of British North America.

"I have, through the whole," Romans begins, "adhered so strictly to truth, as to make no one deviation therefrom willingly, or knowingly." He offers detailed descriptions of the "climate, soil, water, general productions of the earth, the inhabitants with their Customs and manners, and the diseases incident to the human race here." Following a vivid description of the 1772 hurricane, Romans lists Florida flora, speculates on the ethnic origins of the natives (with slightly prurient accounts of their sexual behavior), gives moral justification to slavery, devotes a long section to local cultivation of indigo and madder (previously published in the *Royal American Magazine*), offers an encouraging economic balance sheet for immigrants from New England and the middle colonies— passage money, provisions, tools, livestock, official fees, home construction, slaves and initial living expenses requiring only $2,500 capital— and runs through a terrifying catalog of tropical diseases.

Romans also prints the lengthy daily journal of his West Florida exploration from September 1771 to January 1772— that tantalizingly suggests similar autobiographical material from later periods of his life may have existed and since disappeared. A detailed appendix of sailing directions that made the book so popular with seafarers is followed by this last apology: "At the first planning of this publication,

A CONCISE

NATURAL HISTORY

OF

Eaſt and Weſt FLORIDA;

CONTAINING

An Account of the natural Produce of all the Southern
Part of BRITISH AMERICA, in the three
Kingdoms of Nature, particularly the Animal and
Vegetable.

LIKEWISE,

The artificial Produce now raiſed, or poſſible to be raiſed,
and manufactured there, with ſome commercial and political Obſervations in that part of the world; and a chorographical Account of the ſame.

To which is added, by Way of Appendix,

Plain and eaſy Directions to Navigators over the Bank of
Bahama, the Coaſt of the two Floridas, the North of
Cuba, and the dangerous Gulph Paſſage. Noting alſo,
the hitherto unknown watering Places in that Part of
America, intended principally for the Uſe of ſuch Veſ-
ſels as may be ſo unfortunate as to be directed by
Weather in that difficult Part of the World.

By Captain BERNARD ROMANS.

Illuſtrated with twelve COPPER PLATES,
And Two whole Sheet MAPS.

VOL. I.

NEW-YORK:
Printed for the AUTHOR, M,DCC,LXXV.

9. Rivington's title page for *A Concise
Natural History*, etc. 1775.

[i]

LIST

OF

SUBSCRIBERS

TO THIS

WORK.

A

MR. Benjamin Andrews, Boſton,
Capt. Samuel Andrews, Newbury-Port,
John Antill, Eſq; New-York,
Capt. Vincent P. Ahfield, ditto,
Mr. Thomas Aylwin, Boſton,
Mr. Thomas Allen, New-London.

B.

MR. Theophilact Bache, New-York,
Mr. Iſaac Beers, New-Haven, Charles
b

8. List of subscribers to *A Concise Natural History*, etc. .

it was intended only to be a single volume, not exceeding 300 pages; but at the request of some friends, I have subjoined so many articles, that it swelled imperceptibly to about 800 pages, which made it necessary to print it in two volumes; and as some unexpected accidents have occasioned delays, I will therefore, to atone in some measure to those kind Gentlemen who favoured me with their subscriptions to the maps, deliver them the second volume gratis, as soon as it is published: It is now in the press." Troubled revolutionary times would soon make that an empty promise.

A dozen years later, fellow American Philosophical Society member Ebenezer Hazard, writing to a friend, would refer "bye the bye" to Romans' book and his whole sheet maps as a "paltry, catchpenny performance. From a personal acquaintance with the man," Hazard wrote to Jeremy Belknap, founder of the Massachusetts Historical Society, "I have not confidence enough in his information to think his *History* worth reading." But Hazard's view of Romans' work remains a minority opinion, probably colored by later events of the Revolution. The *Concise Natural History* is still very readable, and Romans' Florida maps— allowing for two centuries of shifting sands— are amazingly accurate.

Romans was now 55— in those days a ripe old age— and still single [with no further mention of a son old enough to accompany his father around the world]. Having become a respected colonial author, Romans decided to forsake the noisy political turmoil of New York City for the quiet of Wethersfield— oldest settlement in Connecticut colony, on a bend in the Connecticut River four miles south of Hartford. He settled down as one of the village's most distinguished residents.

Romans now pressed ahead with his "work of a much more extensive Nature," the history of the Netherlands, firmly promised to readers through ads in the *Massachusetts Gazette.* His life became uncharacteristically quiet, and a

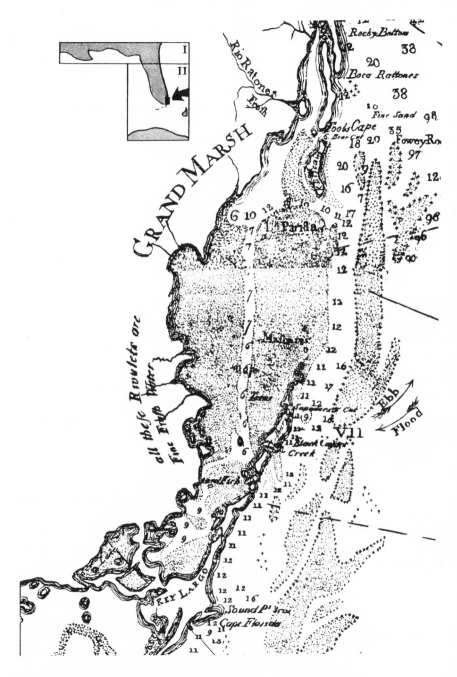

10. Key Largo (and present Miami) area, actual size from 27½ sq. ft. Sheet II [1:600,000], Romans' *Part of the Province of East Florida* 1774, engraved by Paul Revere.

local 16-year-old, Elizabeth Whiting, caught the older man's eye.

But then Bernard Romans, too, heard the shots from Lexington Green and Concord Bridge and the counterpoint of that different American Revolutionary drum. Old relationships were swiftly turning upside down; only a few miles away from Wethersfield, the Connecticut Assembly was drafting its grim picture of those disasters awaiting colonies who failed to immediately respond to the unforgivable abuses of Parliament: "Religion, property, personal safety, learning, arts, public and private virtue, social happiness and every blessing attendant upon liberty will fall victims to the principles and measures advanced and pursued against us, whilst shameless vice, infidelity, irreligion, abject dependence, ignorance, superstitition, meanness, servility and the whole train of despotism, present themselves to our view in melancholy prospect."

The polemic did not fall on deaf ears. Romans may have felt himself ready for a different and more responsive American social system, to replace the one under which he had struggled with insufficient recognition for 35 years. He was hardly a patriot youngster (Washington was 12 years his junior; Hamilton and Lafayette were then only boys of 18). Yet the older man could see glory ahead in the looming conflict— particularly for someone who had long harbored military ambitions. Romans' personal involvement in the coming rebellion would surely bring an end to his small but useful Crown "pension," but he still found himself ready and willing to take up arms in the patriot cause.

Romans ignored the hastily circulated, dire conservative warnings that any further violent action might lead to throwing off "all subordination and connexion with Great Britain," and with "every fiction, falsehood and fraud, delude the people from their due allegiance, throw the subsisting governments into anarchy, incite the ignorant and vulgar to arms, and with those arms establish American independence."

Or perhaps it was a retaliatory streak that led Bernard Romans away from the simple peace of Wethersfield Common to risk his life in the struggle for America's independence? To replace his vanished dream of a royally-sponsored trip around the world, the middle-aged Americanized Dutchman sought a replacement in an equally dramatic revolutionary enterprise.

11. First page of Romans' £131 11s. 10d. Ticonderoga/Fort George expense account drawn on Connecticut Colony.

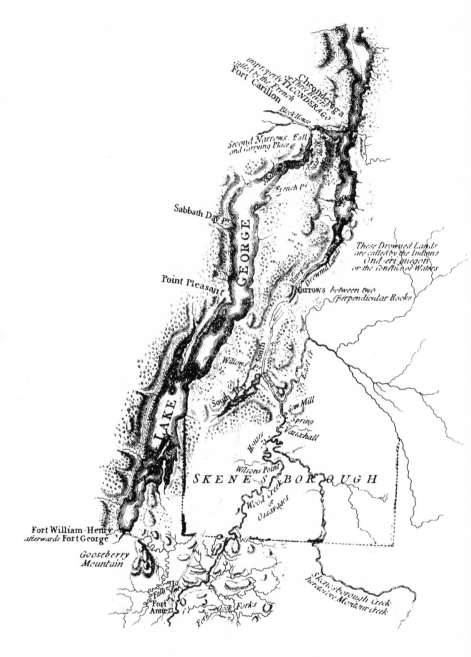

Chevelogera
improperly or Thine River
called by the French TICONDEROGO
Fort Carillon

Block House

Second Narrows Fall
and carrying Place

French Pt

Sabbath Day Pt

GEORGE

These Drowned Lands
are called by the Indians
Ond-eri queqon
or the confluence of Waters

Point Pleasant

Narrows between two
perpendicular Rocks

River Drowned Lands

LAKE

Willows
Pt Smith

East Cr

South Cr

Saw Mill
Spring
Vauxhall

Wilsons Point

SKENES BOROUGH

Fort William-Henry
afterwards Fort George

Wood Creek
or Ossawages

Gooseberry
Mountain

Fall
Fort
Anne

Forks

Skittesborough Creek
heretofore Montour Creek

12. *Lake George and Fort Ticonderoga*: William Brassier's map
1762 (engraved by Sayer & Bennet 1775).

CHAPTER II

Ticonderoga

*In which Bernard Romans patriotically answers
his country's call, marches off to capture a
great fort in the wilderness . . . and loses his
initial command. Bouncing back, he makes the
first engraving of the Battle of Bunker Hill.
Meanwhile, Congress decides to fortify the Hudson River.*

Less than a week after hastily-gathered Massachusetts
Minute Men had driven Pitcairn's and Percy's British
regulars back from Concord Bridge to the protection of the
Royal Navy cannon on the Charles River, Bernard Romans
rode up into Hartford for a fateful interview with the newly-
formed Connecticut Committee of Safety. That revolutionary group included Romans' neighbor, the successful
Wethersfield merchant and land speculator Silas Deane,
ready to depart as one of the colony's delegates to the
Second Continental Congress in Philadelphia; Samuel
Holden Parsons, soon to become one of Washington's
more reliable generals; and Samuel Wyllys, son of the long-
time Secretary of the Colony. Without timidity or indecision,
Romans suggested to the Safety Committee a dramatic
military operation to succor an army woefully lacking in
heavy artillery.

It was "an expedition of the utmost importance," the
editor of London's *Annual Register* complained anony-
mously a year later, "which not only in its consequences
most materially affected the interest and power of government in the colonies; but brought the question to depend
merely upon accident whether we should have a single
possession left in North-America."

Romans' plan was based on a swift march 150 miles north
through the New England woods to seize the undergarrisoned

49

fortifications at Ticonderoga (the native American "place between the big waters") and Crown Point, before those posts could be reinforced by a British expedition from Canada. Capture of the forts, Romans insisted, could eventually lead to delivery of all their cannon and military stores to the American army around Boston.

The proposed raid was not a unique idea; a few weeks before, the patriot John Brown had counseled from Canada: "One thing I must mention to be kept as a profound Secret, the Fort at Ticonderoga must be seized as soon as possible should hostilities be committed by the King's troops." But such an irrevocable act of patriot defiance required a courageous leader with the tenacity to carry it out. To that purpose, Benedict Arnold, former trader to Montreal and now Captain of the New Haven Company of the Connecticut Governor's Foot Guards, had taken advantage of a recent Canadian trip to spy out Ticonderoga's artillery. Arnold's inventory: "Eight pieces of heavy cannon, 20 brass guns— from four- to eighteen-pounders— and ten to 12 large mortars."

Despite the desire to have such patriot interest "kept as a profound Secret," this unmistakable attention to a frontier post hundreds of miles from beleaguered Boston was not lost on Ticonderoga's commander, Captain William Delaplace. He wrote nervously to General Gage, the North American commander-in chief, now ending his first year as joint civilian and military governor of Massachusetts. Gage replied to Delaplace's alarum on March 8th, and urged him to somehow guard against any surprise attack on the isolated and inadequately garrisoned fortification.

Both British captain and general had cause for concern. "The place," Arnold told Connecticut Safety Commissioner Parsons as their paths crossed on the New Haven-Cambridge pike a few days before Romans' interview, "could not hold out an hour under vigorous attack." Now ambitious Captain Arnold was racing east to the patriot lines at Boston, to tell a similar story to Dr. Joseph Warren and the Massachusetts Committee of Safety. Arnold urged that he be ordered north immediately on "secret service" with a Massachusetts colonel's commission in his pocket— to replace his lower Connecticut rank.

The Massachusetts Committee, observing Fort Ticonderoga's location in New York, voted to forward Arnold's suggestion to that sister colony. In Hartford, however, Deane, Parsons and Wyllys cared less about the geographic niceties. They were quick to move, before the Second Congress— ready to meet in Philadelphia with a disturbing number of members vocal for swift reconciliation with Great Britain— could possibly disavow such an act of reckless aggression.

The ferment grew. On April 27, 1775, the day after the Connecticut Assembly ordered six regiments of one thousand militiamen each be raised to support the Massachusetts troops around Boston, the Committee of Safety started their Wethersfield volunteer Romans north from Hartford with extralegal revolutionary orders to take possession of "Ticonderoga and its dependencies." They gave him the rank of captain and "£100 in cash" to raise troops en route. The following day they hurried along a more experienced Captain Edward Mott with five other soldiers to catch up with their newly-minted officer— in such haste that Romans later had to advance Mott 50 shillings to buy himself a gun.

Immortality perfumed the cool spring air. On the eve of the May 10th attack on Ticonderoga, the original Hartford expedition of seven men had swelled to more than 200 volunteers from various parts of New England and New York. But as the group worked its way north, Romans' scheme came unstuck; there was disagreement over the ability of an elderly Southern engineer to lead such a large body of hotheaded Yankees. To add to Romans' problems, the expedition had hardly crossed over into Vermont— the "Hampshire Grants" claimed by New York Colony— when two more ambitious officers caught up. They were Colonel

Benedict Arnold, flourishing a provisional Massachusetts commission from Dr. Warren, and the Green Mountain firebrand Colonel Ethan Allen.

Both of these senior officers, squabbling among themselves, asserted command of Romans' original expedition. "Colonel Allen," protested Arnold, "is a proper man to head his own wild people, but entirely unacquainted with military service." But Allen's rough-and-tumble boys held the majority. Arnold worked out an uneasy accomodation with Allen; Romans was completely ignored and swiftly deposed.

Connecticut Captain Mott's journal tersely relates how, at Bennington's Catamount Tavern, "Mr. Romans left us and joined no more; we were all glad, as he had been a trouble to us, all the time he was with us."

A lesser man, thus disowned, might have crept miserably home. But not Bernard Romans. He had marched north for a taste of glory, and even singlehandedly he was determined to possess it— stepping to whatever drum he heard, however measured or far away. Without delay, Romans set upon an alternative military post that, by some stretch of his orders from the Connecticut Committee, could be considered a "dependency" of Fort Ticonderoga. It was old Fort George, built in 1759 to replace Fort William Henry at the foot of Lake George.

A relic of the French and Indian War, Fort George had been all but abandoned for eight years, and was at the moment in the care of 65-year-old retired British Army Captain John Nordberg, and a single aide. Nordberg was a "half pay officer invalid, twice shot through my body and pleagd with Gravell." He had arrived in New York City from London only half a year earlier, and had "heard nothing els than disharmony amongst Gentlemen, which was not agreeable to my age." With royal "liberty to live where I please in America," Nordberg had travelled north to the dilapidated fort on the beautiful lake, and dwelt "very happy in a little Cottage as an hermit." On a military allowance of "7 shilling sterling per day," his job was to keep an eye on the fort and assist "any express going between New York and Canada."

With all the attention that had been paid to Ticonderoga, one must assume everyone, including Romans, was also aware of the dismal state of Fort George. Nevertheless, marching 50 miles through the woods— picking up 16 new volunteers ("expenses £1 10s.") en route— Captain Romans presented himself at the ruined gate of the fort, dismissed the aide, arrested Nordberg, and sent him off on parole to the patriot base at New Lebanon NY, in what the ailing officer later told the New York Provincial Congress was a "very genteel and civil" manner.

Having snatched this trifling but demonstrably honorable reward from the jaws of disgrace (in the same week that Whitehall was approving a further continuance of his £50 West Florida botanist salary for 1774-1775) Romans' ego was sufficiently restored to permit him to rejoin his victorious former comrades at Fort Ticonderoga on May 14th. Swallowing his pride, he helped to inventory the captured weapons, making a salutary impression on Colonel Arnold, 21 years his junior, who observed to the Massachusetts Committee of Safety that Romans was "a very spirited, judicious gentleman, who has the service of the country much at heart"; a man (Arnold said somewhat smugly, after having usurped his command) whom he hoped would "meet proper encouragement."

To the expedition's enormous satisfaction, Ticonderoga's artillery and military stores— husbanded by the Crown in the wilderness since 1759— were loudly and properly hailed as the new "armory of the Revolution." The materiel far exceeded Arnold's original estimate. Various parts of the works held 78 cannon in good condition, six mortars (three weighing a ton apiece), three howitzers, thousands of cannonballs, 30,000 musket flints and a ton of lead. Plans were laid to transport the heaviest guns by ox-drawn sledge to Boston, once New England's rivers iced over.

Meanwhile, Colonel Allen paid the unfortunate commander Captain Delaplace £18 11s. 9d. for "Ninety Gallons of Rum of his own property, appropriated for the use" of the new garrison; the Green Mountain Boys made merry. The dilapidated condition of the fort itself, key to the land-and-water military path between New York and Montreal, presented a different problem. Arnold wrote to the Massachusetts Committee of Safety that the post was "next to impossible to repair without 1,500 immediate reinforcements. I have the pleasure," he added, "of being joined in that sentiment by Mr [sic] Romans, who is esteemed an able engineer." Arnold dispatched his new aide southward to seek "carriages for the cannons, &c., and provisions."

By then, Congress had caught up with the rapid escalation of military events. After bitter wrangling, the Philadelphia legislators temporarily outvoted their active New England delegation— as the Connecticut Committee of Safety had correctly feared they might— and passed a resolution that counselled a pullback from the captured forts in the northern wilderness. Arnold's and Romans' careful inventory soon became the basis of a projected "safe return" of His Majesty's property once "the restoration of the former harmony between Great Britain and these colonies so ardently wished for by the latter shall render it prudent."

At Albany, Romans attended to Arnold's assignments as best he could. Then, considering his Ticonderoga tour of duty complete, he set off for home. In Hartford on May 31st he delivered an itemized accounting of his £100 advance, plus additional expenses of £31 11s. 10d.— paid to him by the colony treasurer without delay.

Welcomed as a local hero in Wethersfield, Romans learned of Congress' military and political indecision with resentment. In a matter of weeks, however, more important news came from Boston, where in new and heavy fighting, the Massachusetts commanders had applied brief but dramatic counterpressure to a British plan to fortify dominating heights on the Dorchester and Charlestown peninsulas. To Romans, it was a golden opportunity for topical mapmaking— and perhaps even grander graphic projects; during the Revolution, maps and pictures of the "seat of war" were quite saleable. The times were chaotic; the engineer's opportunities for making money were rapidly disappearing; the costs involved in issuing his *Concise Natural History* had swallowed up the subscription fees; inevitably, news of Romans' military actions would reach Whitehall and bring an abrupt end to his £50 "pension."

Unforgivable, too, was the way in which Allen and Arnold had successfully cheated him of his first revolutionary command. Romans' response was to pack his pencil case, drawing instruments and paper and head east along the Boston pike— to where things were again happening.

          *   *   *

Two days before Nordberg's *opéra bouffe* capitulation of Fort George, the Second Continental Congress convened in Philadelphia. Main order of business was the precarious situation of the colonies. Principal military concern focussed on the British army in Boston. However, based on New York's experiences in the strategy of past colonial wars, that colony's members forcefully called Congressional attention to the danger to the entire patriot cause from a fresh British expeditionary force moving against New York.

"As the enemy gains knowledge of the country," said New York, "they must be more and more convinced of the necessity of becoming masters of the Hudson River. It will give them the entire command of water communications with the Indian nations, effectually prevent all intercourse between our eastern and southern confederacy, divide our strength and enfeeble every effort for our common preservation and security."

That warning was timely. Some impregnable patriot barrier to block the Hudson became a national, not a local necessity. For their part, the British also recognized the strategic importance of the river. It was the easiest path between the southern and northern parts of New York colony; and its western tributary, the Mohawk, provided a gateway from the seaboard to the Great Lakes area.

In London, the War Office had already considered —in addition to prompt seizure of New York City— stationing a number of small men-of-war and naval cutters along the Hudson. Such vessels, they reasoned, would "provide safe intercourse and correspondence between Quebec, Albany and New York, and in conjunction with the Indians, allow continual irruptions into New Hampshire, Massachusetts and Connecticut, and so distract and divide the Provincial forces as to render it easy for the British Army at Boston to defeat them, break the spirits of the Massachusetts people and compel an absolute subjection to Great Britain."

Commented Virginia delegate George Washington, who had now taken to wearing his old French and Indian War uniform to sessions of the Congress: "The importance of the Hudson river in the present contest and the necessity of defending it, are subjects which have been so frequently and fully discussed and are so well understood, that it is unnecessary to enlarge upon them. It is the only passage by which the enemy from any part of the coast can ever hope to cooperate with an army from Canada. And further, upon its security, in a great measure, depend our chief supplies of flour for the subsistence of such forces as we may have occasion for in the course of the war."

The patriot choice was a permanent one: defend the Hudson or abandon it. The "advance when possible, retreat when pressed" strategy that served Washington so well elsewhere along the Atlantic seaboard could never apply to the Hudson River. For both armies, until the intervention of the French fleet, the Hudson was the only fixed line of operations of the Revolution.

13. "Northern gate" of the Hudson Highlands.

Moving to secure this strategic waterway from interdiction, spurred by the mounting threat of a "cruel invasion from the province of Quebec," the Continental Congress, sitting as a Committee of the Whole on May 25th, unanimously recommended to New York— as part of its general Declaration of War against Great Britain— "That a post be taken in the Highlands, on each side of the Hudson river, and batteries erected" to block with cannon fire any attempted passage of the river. Following the seizure of Ticonderoga, this was the second important gesture of armed resistance to the Crown in New York Province.

(The river battery idea was sound: five years later, Col. James Livingston, commanding at King's Ferry [Verplanck], became incensed at the British spy-sloop *Vulture's* dawdling in Haverstraw Bay for four days, awaiting the return of the spy Major John André. Livingston brought ammunition several miles from West Point for a single four-pound gun which he hauled to the bank of the Hudson at Teller's [Croton] Point— where the river was far wider and presumably safer for naval operations than anywhere in the Highlands. Within an hour and a half, the impetuous infantry colonel had succeeded in hulling the British warship six times; its captain, Andrew Sutherland, suffered a bloodied nose. The *Vulture* was forced to slip its cable and drop downstream to Ossining. André was deprived of his water transportation home, and Arnold's plotted treason began to unravel.)

Receiving that resolution from Philadelphia to "take a post," the New York Provincial Congress established a committee that included Colonel James Clinton and Christopher Tappen, both Highlands residents (and brothers-in-law), to visit the area and recommend "the most proper place for erecting one or more fortifications."

What exactly are the Hudson "Highlands"? Thirty miles north of New York City, a ten-mile-wide band of Pre-Cambrian granite forms a series of rugged hilltops that rise 1,400 feet above the Hudson's glaciated sea-fiord. Timothy

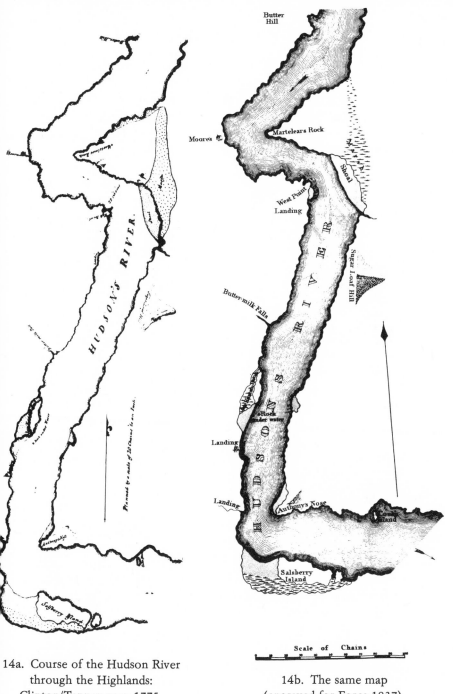

14a. Course of the Hudson River through the Highlands: Clinton/Tappen map 1775.

14b. The same map (engraved for Force 1837).

Dwight, future president of Yale (and one of the youngest subscribers to Romans' *Concise Natural History*) served in the area as an army chaplain during the Revolution. He found the Highlands "majestic, solemn, wild and melancholy." In 1775, their lower summits as well as the high banks of the river offered superb artillery positions against any British naval force sailing north.

By the second year of the Revolution, to control far-flung patriot military operations over a million-square-mile area that reached from Georgia to Canada and as far west as the Mississippi, Congress established six territorial departments. From late 1776 to the end of the war, the tiny but critical Highlands area— less than 150,000 acres— was considered as a very important de facto seventh department.

Within two weeks Clinton and Tappen, drawing a map that considerably distorted the actual course of the river, recommended construction of a major fort near the northern "gate" of the Highlands. It was at the deepest point of the Hudson (175 feet deep), at a place where the estuary narrows to less than a quarter-mile wide, and hampers navigation with a difficult S-shaped bend. For reasons difficult to comprehend, the two-man committee suggested building a post on the lower eastern bank, rather than on the higher western bank of the river. It was a small decision taken to meet the needs of the moment; from it would arise all manner of unforeseen consequences.

Next step for the Provincial Congress was to find a military engineer capable of supervising construction of the fortification— at a cost estimated around £1,500. But New York Province's search for such a person yielded nothing but frustration; the few engineers available along the seaboard were busy elsewhere, hastily reconstructing defenses that had languished since the end of the Seven Years'— French and Indian— War.

The chaos was understandable; the Revolution was still sorting itself out. At one and the same time, it was being fought by independent individuals, military units and government committees. Any knowledge of fortification technology that existed among a tiny handful of patriot officers was based mainly on foreign military texts. (Colonel Rufus Putnam devised his successful patriot "entrenchments" on Dorchester's hard-frozen heights, ending the occupation of Boston, from a description he found in a book by Clairac that he accidentally came across in General William Heath's tent. Putnam and books went together; he had even subscribed for six copies of Romans' *Concise Natural History.*)

Most of the subsequent engineer-volunteers of the Revolution came from European armies. Louis le Begue de Presle Duportail, who later served as Chief Engineer of the Continental Army for five years, was a French Royal Engineer; his classic training in siege warfare was not decisively exercised until the final major battle at Yorktown. Congress would not authorize a formal Corps of Engineers for almost another four years, even though on July 10th, 1775, Washington was commenting bitterly to Continental Congress President Hancock: "The Skill of those [engineers] we have is very imperfect, and confined to the mere manual exercise of cannon; whereas the war in which we are engaged requires a knowledge of fortification. If any persons thus qualified are to be found in the southern colonies, it would be of great public service to forward them with all expedition."

But no such construction engineers were forthcoming, so the various government bodies may be excused for dragging their feet about the desperate military need to "take a post" in the Highlands.

*   *   *

By now, Bernard Romans was back in Connecticut, hard at work on two topical projects spawned by his flying trip to Boston. Again working with engraver Abel Buell, Romans was preparing a two-square-foot "complete and elegant Map showing the Seat of the present unhappy war in North-America." Diplomatically, he dedicated the map to *Concise Natural History* subscriber John Hancock and even indicated the Congress president's enemy-occupied house on Boston Neck.

For the map, "Author, Bernard Romans," now a veteran of the mythic capture of Ticonderoga, burst out with full Revolutionary lyricism: "HAIL, O LIBERTY! Thou glorious, thou inestimable Blessing! Banished from almost every Part of the Old World, America, thy Darling, received thee as her Beloved: Her Arms shall protect thee— Her Sons will cherish thee!"

Romans' cartography— covering a wide area around Boston from Providence RI to Salem MA— was priced at 5s. (coloring extra), and heavily advertised in newspapers as coming from the hand of an "able Draughtsman who was on the spot at the engagements of LEXINGTON and BUNKER'S HILL." Romans, however, was present at neither battle. "During the engagements" would have been a lie; "at the engagements" is at best ambiguous. But it was characteristic of the manner in which this "ingenious man" was always able to pitch his tent in the shade of the truth.

In any event, the map was gobbled up. "Every well-wisher to the country," wrote Nicholas Brooks, who printed the map for Romans in Philadelphia, "cannot but delight in seeing the plan of the grounds on which our brave American Army conquered the British Ministerial Forces" (Brooks was also carried away).

Romans' second contribution to the earliest graphic propaganda of the Revolution was a major creative breakthrough for an artist whose previous figurative work on copper had been limited' to awkward cartouches, unusual plants and the heads of Florida natives. Signed *B: Romans in AEre incidit* [in copper engraved] at bottom right, this dramatic 11" x 16" *Exact View of the Late Battle at*

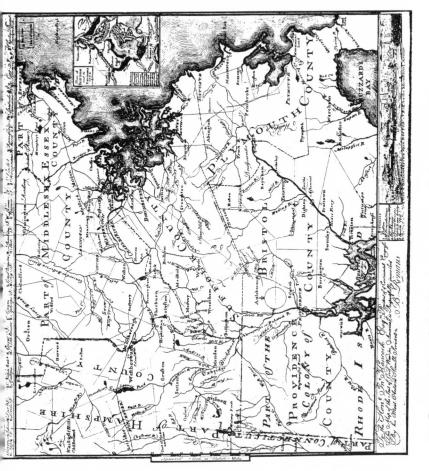

15. Romans' *Seat of Civil War in America* 1775, dedicated to John Hancock.

AN EXACT VIEW of THE LATE BA

In which an advanced party of about 700 Provincials stood an Attack made by 11 Regiments

Leaving Eleven Hundred of the

16. Romans' *An Exact View*

of the Late Battle, etc. 1775.

Charlestown is an idealized and striking panorama of what rapidly became known as the Battle of Bunker Hill.

It portrays the beginnings of the action, with Charlestown afire and the first wave of General Gage's regulars moving stolidly up Breed's Hill against the hastily-constructed patriot fort at the crest. Romans' plate displays a reasonable level of compositional skill (the large tree framing the left side of the picture was a famous Boston Harbor seamark), and some technical ability. The heirloom print provided eager buyers throughout the colonies with a timely report on this first major test of American arms.

A few copies of Romans' map and *Exact View* (also printed in August 1775 by Nicholas Brooks in Philadelphia), still survive [a copy of the *View* would be auctioned 200 years later for three times as much as Romans probably earned in his entire lifetime]. The *View* was also quickly copied half-size by Robert Aitken, another Philadelphia engraver, publisher and bookseller, for the September issue of his successful *Pennsylvania Magazine, or American Monthly Museum* edited by Thomas Paine (later Aitken would also put a second edition of Romans' *Concise Natural History* through the press). In June of the following year, the *View* was re-engraved in London by Wallace & Stonehouse, presumably for the edification of redcoats' widows. It had rapidly become an icon for both sides of the Revolution.

From a military point of view, the summer that followed the bloody action at Bunker Hill seemed uneasily calm. The British lay quietly in Boston, slowly recovering from the shock of taking 1,056 casualties— 40% of their finest North American troops— in a single afternoon. For their part, many Americans still hoped the mounting storm would somehow blow by. As Romans completed tooling his copper plate, word reached as far as Wethersfield that New York was searching— apparently with no great urgency— for a qualified engineer to supervise construction of an important fortification along the Hudson River.

Why not— Romans must have thought— the King's own

eminent Southern District surveyor-cartographer-engineer with 20 years' experience; now the well-known liberator of Fort George, cannon-counter of Ticonderoga and avid student of military history in the Low Countries? Romans, living through a period that rewarded inspired, aggressive improvisation, had certainly earned brave Benedict Arnold's praise.

To Romans, the opportunity was worth a quick trip past New York and down to Philadelphia, to test Congressional sentiment, seeing whether recommendations would be forthcoming from old and new friends. Also, while he was in that buzzing revolutionary capital, he could lay his new graphic material in front of a publisher. Romans packed the plates in his saddlebags for the long 300-mile trip from Wethersfield.

And clouding his vision as he rode southward was an image of a fantastic, impregnable riverine fortress, the likes of which North America had never seen.

MARTELAER'S ROCK (CONSTITUTION ISLAND)

17. Martelaer's Rock (Constitution Island) from the south.

18. Memorial cannon marking the site of Romans' 1775 unusual blockhouse commanding the narrowed Hudson [West Point at right] (1983).

CHAPTER III

Martelaer's Rock

*In which Bernard Romans convinces two
Congresses that he can build them a great patriot
fort along the Hudson, is quickly at odds with the
Commissioners appointed to supervise his work,
and goes behind their backs.*

On August 18, 1775, with the British forces still tied
down in Boston, the New York Provincial Congress took a
brief recess. They temporarily entrusted control of the
Province to a key "Committee of Safety"— but first
resolved, after a summer delay that had already wasted two
months of good construction weather, "That the Fortifications ordered by the Continental Congress as proper to be
built on the banks of Hudson's River be immediately [sic]
erected." To implement their resolution, they appointed
five "Commissioners for Fortifications at the Highlands"
(increased to seven the following month), paid 10s. a day
each.

The Commissioners, with a guard of 24 men, were
established as a quasi-independent group, charged with
recruiting laborers, managing necessary payments and
generally overseeing construction progress on the new
fortifications.

The move by the Provincial Congress also underlined
their serious approach to the Hudson situation. By a vote of
18 to six, they authorized removal upriver of several
cannon— from the Grand Battery that defended the
southern tip of Manhattan Island itself.

Meanwhile, as he had hoped, Bernard Romans found
himself quite welcome in Philadelphia. The situation there,
as Charles Carroll of Carrollton observed later to the

Maryland Council of Safety, was one in which "we must avail ourselves of the skill of such [engineers] as we can meet with, though their Knowledge be not so perfect or complete." Romans had come to the right place at the right time for a project as monumental as his ambition. Buttressed by his membership in the American Philosophical Society, he lobbied the Continental Congress with great success.

For extra brush-up, Romans purchased from publisher Robert Aitken on September 5th a copy of Roger Stevenson's new 232-page *Military Instructions for an Officer Detached in the Field* (plus an *Essay on the Character of Women* and a handsome new silver map-pencil case), and headed north for New York. Upon his arrival in the city, Romans immediately called upon the Commissioners for Fortifications, who accepted the Congressional approval of his qualifications. They recommended to the New York Committee of Safety that Romans be placed in charge of planning and constructing their strategic river installations.

There was one major difficulty; the Committee established no direct lines of responsibility and authority. But "Mr. Romans the Engineer" was hardly one to complain; he was delighted to seize leadership of a project that would surely involve him with some of America's most illustrious military men.

The first site Romans eagerly surveyed, following the original (but still strategically inappropriate) suggestion from Clinton and Tappen, was the rugged 160-acre island called "Martelaer's Rock." The name corrupted *martelaar's reik* (martyr's reach), the Dutch skippers' term for that extremely difficult tack on the river. The island, with its highest point only 150 feet above water level, lay in the midst of the Highlands, across less than a quarter mile of open water from the "West Point." An extensive marsh separated it from the proper east bank of the river.

Clambering along the southern shore of the rock, crossing its few level areas only several paces wide, Romans began to draw upon half-forgotten bits of formal British engineering training, plus whatever fortification experience he had acquired under de Brahm. (The latter, despite a nephew who became a patriot military engineer, remained a staunch loyalist, and was now counseling construction of forts large enough to hold 3,000 men plus a governor's residence, to "separate the garrison and King's officers from the inhabitants" of rebellious North America.)

Faced with the greatest professional challenge— and opportunity—of his lifetime, Romans' imagination took wing.

He proposed to build, with staged payments from the New York Provincial Congress, a very large military complex indeed: an unusual octagonal blockhouse/magazine "to prevent mischief from a vessel's top"; five batteries with 81 guns; a fort with bastions ("the soul of the works"); a 200-foot long curtain rampart; several subterrannean bombproofs for protection and additional magazines; a 100-foot-long barracks; storehouses and guardrooms. Unafraid to improvise, Romans planned 14 cannon for the curtain, "although according to rule it ought to have only ten." With his total of 81 guns, Romans promised "a most terrible crossfire, to make it totally impossible for a vessel to stand it" — even the most heavily-armed British ship able to maneuver up the river.

For this involved fortification plus additional outworks, Romans planned to charge New York Province— he included materials of every description, "Labour of & Provisions for 150 Men 4 months 26 days to the Month at an Average of 3s. pr. day"— £4,645 4s. 4d., all "computed at the lowest rates available." Romans' estimate was certainly precise, covering everything except "150,000 bricks, the price of which I am entirely ignorant of."

If the King's former "principal deputy surveyor" harbored any misgivings about his ability to bring such a grandiose scheme to fruition within the time and funding available, he kept them to himself. His proposed "4 months" of construction, however, should have opened everyone's eyes. In the prevailing military situation, the monumental fort planned was too huge an undertaking for hand tools, blasting powder, a few oxen and an enormous amount of backbreaking human labor. A more practical suggestion, given the country's financial straits, would have been to place a few precious patriot cannon within strategic earthworks thrown up at various narrow turns in the river. Romans' design was much too ambitious for a poor young nation still unable to break the British stranglehold on Boston.

The only formal explanation Romans was able to make for his grandiose scheme was that "a less or more imperfect plan would only be beginning a stronghold for an Enemy." But to the Provincial Congress' Committee of Safety Chairman John Haring, the projected £4,645 cost appeared enormous. As an early example of revenue-sharing, Haring attempted to "save a great expence in labour" by requisitioning a full company of General David Wooster's Connecticut Continentals stationed in New York's Haarlem, to work under Romans. Pending some order to that effect from Philadelphia, the sixty-five-year old Wooster declined.

Less than three months earlier, Wooster had written to Connecticut Governor Jonathan Trumbull about New York's Congress, "I have no faith in their honesty in the cause." Now John Berrien, Commissary Agent for the Fortifications Commissioners, responded by comparing the veteran general of the French and Indian War to "the hand of a Clock; though it moves, the eye cannot discover it." For the next seven months, although Philadelphia was willing to appropriate necessary construction funds, the Province of New York remained responsible for all the work on the fort.

As Romans, paper and new silver pencil case in hand, stalked the uncompromisingly rugged site of his planned Gibraltar, he appeared to ignore two very serious questions. Although the "Grand Bastion" was properly sited on his

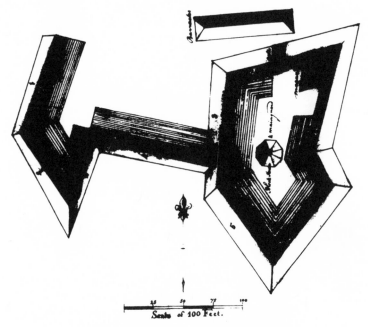

19a. Romans' projected fort on Martelaer's Rock 1775.

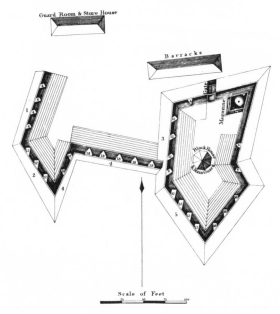

19b. The same drawing (engraved for Force 1837).

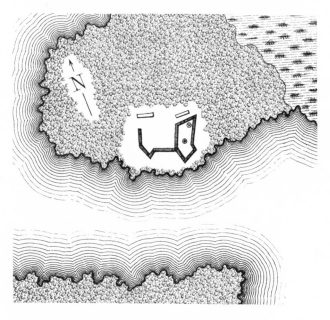

19c. The same drawing (re-engraved for Boynton 1864).

20. Signature page, Romans' letter of transmittal for his fortification drawings.

21. Looking north above West Point to Constitution Island (Martelaer's Rock) 1978.
F = Ruins of Fort Constitution; G = Gravel hill battery site.

somewhat distorted sketch map of the river's S-bend at Martelaer's Rock, in actuality the work was located too far west to command a long straight stretch of the lower river. Until the moment before beginning its own cannonade, an enemy vessel would be masked from the bastion's guns by the West Point— something Romans himself acknowledged (see the upper right of *Figure 25: "Here ought to be 4 heavy cannon or 6 to prevent a Bombvessel Laying at A"*). Inevitably Romans' works would either have to be moved or extended.

The Commissioners came up to the Rock on September 13th to review construction plans. The sightline problems were immediately apparent. Romans had also considered siting a more strategically-placed battery on an easy-to-level (but far less dramatic) gravel hill to the east— but never worked on it! Perhaps the engineer considered that site as mere bait, to lure enemy ships under a devastating fire from his huge fort.

The more basic problem was that even the highest point on Martelaer's Rock was still lower than the opposite shore— beneath any cannon an enemy might some day haul there (Romans' "Grand Bastion" was actually 500 feet beneath the future site of West Point's Fort Putnam). Conflict immediately arose between Engineer Romans and the Fortifications Commissioners.

The focus of their dispute reflected the American Revolution (and probably all revolutions) in microcosm— high vs. low motivation, sacrifice vs. egotism, organization vs. chaos.

On September 18th, deliberately snubbing the authority of the disapproving Commissioners, Romans slipped away from Martelaer's Rock. Talking of a quick trip to Wethersfield, he left the Commissioners behind and hurried south instead, to exhibit his "most frugal plan" of the works and "*Estimate of Expences*" directly to Chairman Haring and the Committee of Safety in New York City. "I most humbly beg this honourable house to pardon the coarseness of the drawings," Romans told the assembled Committee, "they being done in an inconvenient place & at a distance from my Instruments." He then was excused by the Committee to work out a formal contract.

Agent John Berrien quickly transmitted the Committee's initial reaction back to Martelaer's Rock. "On the Whole they seem Pleased," he wrote, "& I must Confess to you I have a high Opinion of the Plan & of Romans' Abilities. They sent off an Express with the plan & Expences to the Continental Congress" (where it was presented to that body September 22nd). The engraver of *An Exact View of Bunker Hill,* despite carefully assumed modesty, had sketched some very convincing military graphics (handsomely copied 60 years later by archivist Peter Force).

That same day, the Committee also sent a pro-forma letter to Beverley Robinson, Esq. at his Highlands estate, two miles south of the proposed fortification. (Later in the war, Robinson's house would alternately serve as head-quarters for both Washington and Benedict Arnold.) Stressing that "the Provincial Congress by no means intend to invade private property," but citing "the necessities of the present times," the Committee asked Robinson to sell them Martelaer's Rock.

The Highlands landlord replied that he didn't own the rock. (It was part of *Lot No. 2* of the huge 1697 Adolph Philipse patent— three-quarters of which Philipse, with the connivance of the corrupt governor Benjamin Fletcher, had pirated from the Muhheakunnuk Wappinger tribe. *Lot No. 2* had by this time passed to Robinson's widowed sister-in-law, Margaret Philipse Ogilvie.) The property actually owned by Robinson and his wife, the former Susannah Philipse, was *Lot No. 1,* opposite Popolopen Creek. But had he owned the rock, Robinson professed, "the publick should be extremely welcome to it." The Committee of Safety decided to eschew further legalisms.

(One may distrust Robinson's apparent generosity. Within the year, against the advice of his friend and neighbor John Jay, he would join the Tory faction, forfeit his Highlands properties and become a colonel in the British Army. In 1777 he led the 400-man Loyalist wing of General Henry Clinton's bloody foray through the Highlands. Three years later, he ran Major Andre"'s ill-fated mission from the cabin of the *Vulture* off Haverstraw.)

On Martelaer's Rock, Fortifications Commissioners Samuel Bayard (lately of the Marine Society's council), William Bedlow and John Hanson were still ostensibly loyal subjects of George III, Whiggishly struggling for America's rights under the Mother Country's cherished— but intangible— Constitution (a declaration of American independence was still a year away). Following Romans' clandestine departure, the Commissioners gathered and formally christened what they were all about to build— *"Fort Constitution."* The patriots' political position was "constitutional," the Tory role "ministerial." The rock, surrounded by river and marsh, has been *Constitution Island* ever since.

On September 25th the bypassed Commissioners learned with indignation (from John Berrien's letter) that the absent Romans had actually appeared before the Committee of Safety in New York, and seen his plans approved and passed to the Continental Congress in Philadelphia without the benefit of their own comments. "We should have esteemed ourselves happy, had we been consulted on this subject before it was sent forward," they bristled to Chairman Haring, "It was easy for one of our body to have waited upon the Committee, to have given them full satisfaction relative to the situation of the ground, which it is not possible for them know by the plan. If we are right in our conjecture, Mr. Romans' plan is not sufficient— it will be only a temporary expedient. Should the fortification fall into the hands of the Ministerial troops, it will prove the ruin of the Province.

"It was not possible for him to have given you any calculation relative to the whole expense," the angry

Commissioners continued, "as it will be absolutely necessary to extend the works. It appears to us that it would have been much better to have calculated the amount of what it would cost, than be obliged hereafter to apply a second time to the Continental Congress."

The Commissioners also addressed what lay at the heart of the matter, the problem of split responsibility: "As we will not be answerable for measures we cannot conduct, we therefore request the favor of you, gentlemen, to inform us whether we are under Mr. Romans' direction, or whether he is obliged to consult with us upon the measures to be pursued. You cannot blame us for this request, as the safety, honour and interest of our Country, and its future welfare, depend upon this important post."

The fat was now in the fire; for the next three months, the irritating question of Bernard Romans' accountability would plague the engineer, and everyone with whom he dealt. That conflict, based on egotism, power and ignorance, hardly considered the needs of the new nation nor the human lives at stake.

To buttress the final observation of their letter, the Commissioners reported that two days prior, Royal Governor Tryon with two military officers— still circulating freely within the inflamed province— had come upriver to Haverstraw, less than 15 miles below Fort Constitution. There the Governor had inquired "with great scrutiny about the new fortification, the nature of the ground, the state it was in, how many guns were mounted and how many men watched."

Meanwhile, in New York, Engineer Romans continued his wooing of the Safety Committee. At 9 o'clock in the morning on the day following their receipt of the Commissioners' angry letter, "Mr. Bernard Romans, attending at the door, was admitted" and submitted his contract to New York Province to cover construction of the installations he had already so convincingly sketched.

To Romans, his best defense against the unwelcome

assertion of authority by the Fortifications Commissioners was to plunge straight ahead. From the Committee, he requested the title of "Provincial Engineer," with "the whole work be done by me for £5,000, the ordnance only excepted; that I may have the management under my direction, and the work already done accepted at the rate expressed in my own estimate; and tools already purchased at the cost; that the Commissioners only have the trouble of supervising my execution, and answering the orders I draw from time to time in favour of the workmen and furnishers of materials; and that £150 or £200 be advanced, to be applied to such incidental matters as are immediately wanted, and do not occur directly to the memory."

The Committee listened silently to Romans' uncompromising proposals, probably a bit embarrassed by the reprimand from their Commissioners. They "Ordered, The consideration be postponed till to-morrow."

The following morning, again at 9 a.m. sharp— the Committee of Safety kept exemplary revolutionary hours— Romans, "attending at the door, was called in." A cloud had gathered. After some discussion of rates of pay, "the committee conversed with him on the subject of his being an engineer at the fortifications, and Mr. Romans withdrew.

"After some time spent thereon, Mr. Romans was again called in, and the Chairman informed him that the Committee would not contract with him for building the said Fortifications; that he should be paid for his services as an engineer; that as the Continental Congress was sitting and the Provincial Congress was to meet in a few days, the Committee could not make any proposal of, or any answer to, establishing him as an Engineer during the unhappy controversy in America.

"But," Chairman Haring continued coolly, "if his merits in the present business should appear to be such as to recommend him, it would be an advantage to his reputation. And that should the controversy unfortunately continue long, it was probable that his future services might be wanted." The Committee insisted, however, "that twelve Shillings a day sterling and not twenty Shillings sterling as he alleged, was the value of the pay and perquisites of an Engineer on the British establishment" and "that since the infancy and present circumstances of the Country will not admit of allowing pay equal to that given in older settled States, that the Committee could not encourage him to expect more than the pay of a Colonel in the Continental army, and that only for the present." The pay of a Continental colonel was comparatively high; at that time the rank was held by less than two dozen officers in Congress' new Army.

To cool Romans' obvious ambition, the Committee suggested that any further consideration of rank properly belonged in other hands: "This proposal shall not stand in the way of any better provision for him, if the Continental Congress should think proper to make any better"; and to soften the matter a little, as Romans now apparently "stood in need of cash, he should have an order on the Treasurer of the Congress of this Colony for twenty Pounds on account." Later that day the Committee of Safety also sent off a conciliatory letter to their Commissioners, reminding them "that Mr. Romans was brought to assist in planning and directing the fortifications by your advice and request" and expressed hope that "the work may now be carried on with all your joint wisdom."

Realizing that the Provincial Congress, scheduled back in session the following week, offered him a broader forum than the now apparently lukewarm Committee of Safety, Romans turned away from the rebuff and began ordering boatloads of heavy rough-hewn lumber for his octagonal blockhouse/magazine and cannon platforms, plus a host of other building materials— all charged to the Commissioners. By October 10th, having swiftly hired 27 carpenters, 16 masons, 2 blacksmiths, 59 unskilled laborers, a clerk and a steward, Romans (with his own slave) was back at Fort Constitution, ready to press construction.

Romans would have been the last to acknowledge any air of panicky improvisation, but the available daylight for outdoor work had already dwindled to $11\frac{1}{2}$ from 15 hours. An unidentified friend of the colonies in London had written to Philadelphia that the British were about to move on New York and Albany, disrupting patriot communications; Samuel Chase of Maryland thundered before an anxious Congress, "Recollect the Intelligence on your Table— defend New York— fortify upon Hudsons River." But in the chilly Highlands, it had become dangerously late in the Fall 1775 season for any serious heavy construction.

22. Bernard Romans' signature;
epitome of Dutch ornamental calligraphy.

CHAPTER IV

Fort Constitution

*In which Bernard Romans' request for
officer's rank is rejected, slow construction
on his fort creates heated argument, alternative
sites are considered, and the Commissioners get
into a little trouble of their own.*

Captain Bernard Romans' unsettling experience with
Colonels Allen and Arnold on the march to Ticonderoga
had provided an object lesson in the values of superior
military rank. He waited only two days to begin a ingenuous
campaign to obtain a full Colonel's commission. Twisting
the salary statement from the Committee of Safety, he
wrote the now-reconvened New York Provincial Congress,
affirming that the Committee "gave me their words that I
should be appointed principal Engineer for this Province,
with the rank [sic] and pay of Colonel. As I have been now
actually engaged in this work since the 29th of August last
[the moment he arrived in Philadelphia to lobby the
Continental Congress?], I should be glad to know the
certainty of my appointment, and therefore humbly pray
that my commission be made out and sent."

"I have left the pursuit of my own business, which was
very considerable," Romans went on, "and endangered my
pension from the Crown [the salary contingent on his
continuing to botanize in West Florida], by engaging in our
great and common cause. These matters considered, I hope
my request will be thought reasonable, and therefore
complied with. I remain, with the utmost respect, honourable Gentlemen, your most obedient humble servant, B.
ROMANS."

The Provincial Congress, angling for an engineer, had
hooked a Tartar. At that moment, commissions for

qualified military men were not too difficult to come by, but the idea of naming Romans a colonel apparently raised hackles among many New Yorkers who knew him from past argument. (Eventually New York would duck the problem altogether, and pass his request to Philadelphia.) The Fortifications Commissioners were already complaining that under Romans' direction, laborers at Martelaer's Rock were busy executing "fancy touches at intolerable public expense." With the Revolution hardly under way, the man who had only partially marched on Ticonderoga— the Commissioners asserted angrily— was erecting not a badly-needed fort but some kind of military monument to himself. The quarrel festered.

The Captain from Connecticut, much like a child playing with new blocks, plunged blandly ahead, insisting the Commissioners had no right to criticize his construction. His authority, he argued, derived solely from the May 25th Resolution of the Continental Congress, which he interpreted to mean that New York Province would merely furnish men, materials and money— any comment on how they were employed must wait until all work on Fort Constitution was complete.

The argument raged. Romans, still a civilian, continued to build, but his over-ambitious designs went at a frustrating pace. Heated accusations were traded; the carpenter soon began to blame his tools. News of mounting acrimony reached Philadelphia, where the Continental Congress initiated further discussion of Romans' plans. John Adams' shorthand notes on the debate quote Eliphalet Dyer of Connecticut: "Cant say how far it would have been proper to have gone upon Romains Plan in the Spring [of 1775], but thinks it too late now. There are Places upon that River, that might be thrown up in a few days, that would do. We must go upon some plan that will be expeditious."

Richard Henry Lee of Virginia, with good recall, was willing to back Romans' more elaborate proposals: "Romain says a less or more imperfect Plan would only be beginning a Strong hold for an Enemy." The arguments of Romans' Wethersfield neighbor Silas Deane finally carried the day: "An order went to N. York. They have employed an Engineer. The People and he agree in the Spot and the Plan. Unless We rescind the whole, We should go on. It ought to be done."

As a result, a resolution of Congress again affirmed "That the provincial Convention of New York [restyled name for the colony's legislators] render Hudson's river defensible," and directed the Province to "be particularly attentive to form such works as may be finished before the winter sets in." In deference to skeptical members, however, Congress added, "It is very doubtful whether any stone work can be properly made at this advanced season; it is submitted to the judgment of the Convention whether it could not be more cheaply and expeditiously done by works of wood or fascines."

Doubts had also arisen in Philadelphia as to whether the fort on Martelaer's Rock was being built at the right point on the river; the island had served so long as a wood lot that there was little useable timber left. The Congressional order directed New York "to inquire whether there are not some other places where small Batteries might be erected, so as to annoy the enemy on their passage, particularly a few heavy cannon at or near Moore's house" [a substantial dwelling on the west bank flats above West Point, characterized as *Col. Moore's Folly* on a 1755 map of the river]; "and at a point on the west shore, a little above Verplanck's" [the rocky peninsula eventually famous as Stony Point].

Defense of the Hudson now appeared to be everyone's concern. Jefferson wrote to his brother-in-law Francis Eppes in Virginia: "By means of Hudson's River, [the British] mean to cut off all correspondence between the Northern and Southern colonies." Four days later Major General Philip Schuyler, commander of the Northern Department, wrote Congress President John Hancock

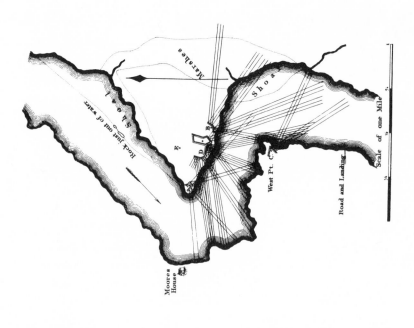

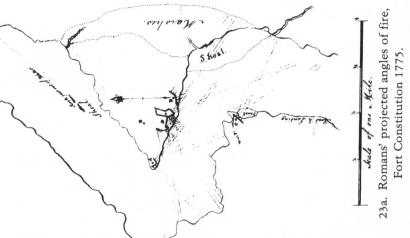

23b. The same drawing (engraved for Force 1837).

23a. Romans' projected angles of fire, Fort Constitution 1775.

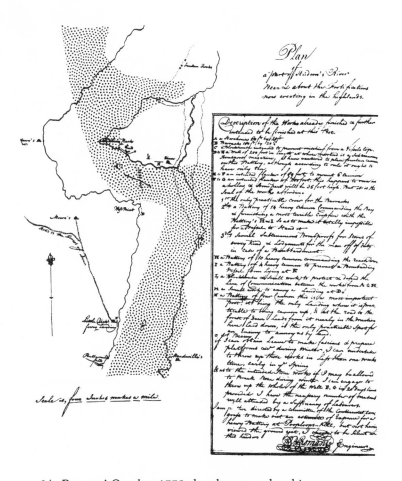

24. Romans' October 1775 sketch map to key his progress report (with true Hudson River course stippled).

from Albany: "To me, Sir, Every Object, as to Importance, sinks almost to Nothing, when put in Competition with the securing of Hudson's River." Romans was certainly performing in a spotlight.

On October 12th, the New York Convention responded to its new instructions from Philadelphia with a letter to their Commissioners, enclosing the text of the order and urging them "without loss of time to go with as much secrecy as the nature of the transaction will admit to the several places mentioned in the resolution, taking Mr. Romans to your assistance, and use all possible dispatch in making your report."

On October 16th, Romans sent an unjustifiably optimistic report on construction progress to the New York Convention, together with his survey of other potential river sites made jointly with the Commissioners. Considering that "the season when this work was undertaken was very far advanced," Romans wrote (with the first appearance of an excuse based on the weather), "I think we are in as forward a situation as can be wished for. I make no doubt that the work will in three weeks' time be of sufficient strength to stand the brunt of as large a ship of rank as can come here, and two or three small fry." He asked New York City for 18 more cannon— 18 to 32 pounders— promising, with no real basis in fact, that they could be ready for mounting on the fortification in about five days.

On what is clearly a direct tracing of the distorted Clinton/Tappen map of the Highlands area prepared several months before, Romans imaginatively catalogued all the "Works already finished or intended to be finished at this Post." Surprisingly— for an experienced cartographer— he copied that map's earlier error of drawing a sharp point on the actually stubby western tip of Martelaer's Rock. Romans' report went on to praise the abundant stone available on the island, compared to the lack of local timber— even fascines for the ramparts were difficult to fabricate: "The part of the work done in timber advances

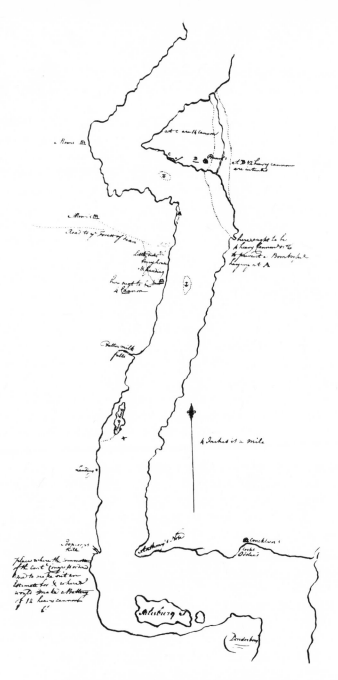

25. Romans' October 1775 alternate site map.

slower than the stone by a degree beyond all comparison," he said, whereas "in four days 150 perches [3,700 cubic feet] has been properly laid by 12 masons."

Tracing out an additional copy of the original Clinton/ Tappen map, Romans proceeded to evaluate the suggested alternate sites. He dismissed the battery suggested by the Continental Congress at Moore's House as "entirely useless," and described "the point on the west side above Verplanck's" as "too easy of access, and in the vicinity of many ill-disposed people." Always looking for increased responsibilities, he had nothing but praise for the possibility of further military construction in the area at "Pooploop's kill [Popolopen Creek], opposite to Anthony's nose. Where [he inscribed on his map with a further invocation of authority] the Committee of the Contl. Congress ordered me to make out an estimate for & where I would make a battery of 12 heavy cannon.

"It is a very important pass," Romans observed, "commanding a great ways up and down, full of counter currents, and subject to constant fall winds; nor is there any anchorage at all except close under the works to be erected." From the Creek, he wrote, "it is a very easy matter to establish posts with the upper country and Connecticut" (to whence, presumably to see his wife, he said he intended "to go in person in about two or three weeks time"). And to demonstrate some rare harmony, Romans asked the Commissioners to append their signatures to his report; a statement by Bayard and Bedlow indicated they "fully concur in opinion with the engineer."

But that momentary truce may actually have been occasioned by a new and different clash of personalities— this time between the Commissioners and several Hudson River skippers. It was a sideshow the uninvolved Romans must have hugely enjoyed, for it proved his opposite numbers had no corner on revolutionary efficiency or good judgment.

At their own level of pomposity, the Commissioners

ordered all traffic on the river to properly identify itself by saluting the new fort a-building. Their hands full tacking though the narrow gut, most sloop captains could not be bothered. Matters came to a boiling head the day before Romans made his progress report: "Since we have had cannon mounted and colours hoisted," the Commissioners explained in a quick letter to Nathaniel Woodhull, President of the New York Convention, "instead of sending their boat on shore, we have thought it necessary for every vessel passing by in the daytime to lower the peak of their mainsail, as a token of their being friendly.

"We are now to acquaint you," their hurried communication continued, "that one Captain Robert North [an unquestioned patriot], passing by on this day, was hailed to lower his peak, which he refused to do. On being threatened with a shot, he replied that was what he wanted. On this we sent an armed boat on board of him, in order to Enquire his reason for refusing to do what he that instant saw another sloop to do.

"On our boat's boarding of him," the Commissioners went on, "he told the people therein he had a brace of pistols, and that if that damned rascal Capt. Bayard did not produce an order to him from the Provincial Congress for the request we made, he would blow his brains out, with many more unfriendly expressions." The incident spelled foolish and unnecessary trouble on the river. Small wonder the Commissioners nervously asked the Congress to signify that "our conduct herein meets with their approbation."

Their letter outsped Captain North to the city by less than a day; he was soon belaboring the New York Congress with his account of "disagreeable treatment received at the new Fortifications." President Woodhull (who less than a year later would die under ambiguous circumstances as a provincial Brigadier-General in the disastrous Battle of Long Island) immediately wrote to Fort Constitution, chiding the Commissioners and suggesting that "you should not of your own authority, without the recommendation of the Congress, exact instances of respect from your fellow-citizens.

"So trivial a token," Woodhull went on, "or the omission of it, can never mark out our friends from our foes, or answer any other valuable purpose. Great disgusts have arisen from your peremptory demand and probably bloodshed may ensue. The colonies have sufficiently suffered through punctilio, and we beg you will desist from exacting marks of submission or respect of any kind, when the peace and safety of the community so loudly forbid it."

Two of the Commissioners at the Fort accepted Woodhull's rebuke; the third, John Hanson, was outraged: "I undertook this business from pure love to my country and to the rights of mankind," he complained, "Never any man was ill-used passing the river, unless he abused us; and then he received a small check that was not equal to his demerits.

"When men act from principle and are placed at any particular post of consequence and find themselves in a precarious situation," Hanson ranted on, "it naturally follows they must exercise their own judgment, and ought to be supported by those who placed them there; which has not been the case in this instance. An insult offered to them was offered to the Congress, and abuse by the complainants ought not to have been permitted. I must therefore request, gentlemen, the favour of the Congress to appoint somebody else in my room, for I will never more go back to the fort."

Two days later the Convention, announcing that "Mr. Jonathan Lawrence of the City of New-York [its member from Queens County] was highly recommended for his great vigilance, activity, care, prudence, skill, management and unremitted industry," appointed him a "Commissioner in the place of Mr. John Hanson, whose private affairs have obliged him to decline that service." (Within seven months, a local Orange County safety committee would accuse

Lawrence and his family of operating a black market in tea with the Fort as their depot.)

On October 27th, after two weeks' mulling over Romans' letter soliciting "the rank and pay of Colonel," the Convention finally wrote to John Hancock in Philadelphia "praying that the Congress will make some Order" regarding the engineer's "Pay of Fifty Dollars per Month for the present," since he had "Objected to them that his Pension from the Crown of fifty pounds sterling per year as Botanist for one of the Floridas might be taken away when it was known that he had Assisted this Country. If Fifty Dollars is thought too little," continued President Woodhull, "will the Congress be pleased to fix upon the Sum which they think adequate to the duty he is to perform?" There was absolutely no mention of a New York colonelcy for Romans.

Meanwhile, with their tempest on a mainpeak now behind them, the reconstituted Fortifications Commissioners again addressed their larger problem— the mounting cost of Bernard Romans' military construction. On November 5th, they wrote once more to Woodhull, this time complaining that "evil persons already amongst us have instilled into the minds of the people at work that there is no security for their pay. This has given us a great deal of trouble with a set of people whose tempers and dispositions are as various as their faces.

"All our influence," the Commissioners went on, "cannot get them to work on Sundays. Some of the artificers employed by the master workman [the Commissioners' new euphemism for Romans] have had the assurance to say they were not to be directed by the commissioners, but by the master workman." Winter was advancing and construction was now far behind schedule. Otherwise, the Commissioners said, "we would have immediately disbanded them from the work."

On November 8th, a special "Committee to the Northward" of the Continental Congress— consisting of Robert

R. Livingston, Jr. of New York, John Langdon of New Hampshire and Robert Treat Paine of Massachusetts (none with any real military experience)— was dispatched from Philadelphia to confer with Northern Department commander General Philip Schuyler, and visit Ticonderoga and Canada. *En passant*, responding to New York's query regarding Romans' pay and other confusion, the committee was charged to "take an accurate view of the state of our Fortifications upon Hudson's River."

Congress wanted a firsthand report on what was really going on. The following day, to underline that concern, the New York delegation in Philadelphia advised the Convention that their fellows had "taken the resolution to appoint a commander of the fortress in the Highlands, with the rank of colonel, a person of abilities, and in whom the inhabitants place confidence." In the event, however, no one was ever appointed.

CHAPTER V

An Exchange of Letters

*In which — while construction continues
to drag — Bernard Romans and the Commissioners
commence a violent paper war, he is termed "a
scoundrel and a villain" (and loses his Crown
"pension"), and Philadelphia sends a committee
to investigate.*

Against mounting difficulties, Bernard Romans continued to work with all the bravado he could muster. His running argument with the Commissioners, reflecting a patriotic but amateurish and poorly-organized approach to a direful crisis, swelled to a crescendo.

The journals of the New York Convention for the first two weeks in November 1775 preserve a memorable series of lengthy and acerb letters that passed between Romans and his opposite numbers: Commissioners Bedlow, Jonathan Lawrence (replacing Hanson) and Thomas Grenell. With a Congressional military appointment hanging fire, all the letters on both sides appear to have been written for the record, with the writers trudging through bitter fall winds between the various temporary shelters on Constitution Island to actually deliver these missives by hand.

The correspondence provides startling witness to what might be viewed as an early American contractor/client style — each party intent on shifting blame and wearing the other down. One is tempted to try to reach back more than 200 years and shake everyone's collar, saying, "Don't you know there's a war on?" Romans' verbiage on this particular occasion also makes one wonder how he, instead of terse Ethan Allen, might have replied to British Captain Jocelyn Feltham's immortal straight line at the Ticonderoga wicket gate: "By whose authority do you intrude here?" Could anything as simple and compelling as "The Great Jehovah and the Continental Congress" have come rolling off Romans' tongue?

Recast verbatim as dialog, Romans accuses the Commissioners of "commencing a paper war. Although I am a great hater of epistolary altercation," he continues, "as in a private station I have more than once exerted myself for America, you may rely on it that I shall do no less, now that I am honored with the post and rank the Congress has conferred on me, the dignity of which commission I shall try to preserve with military vigilance and spirit." But which Congress? What rank?

Juggling the actual number of his workers, Romans then proceeds to criticize the activity of the Fort's 27 masons, supported by 26 laborers, "a distribution the most erroneous that can be imagined." In 28 days, he complains, the group has "only completed 700 perches [17,000 cubic feet] of stone wall," instead of the scheduled 2,400 perches [60,000 cubic feet]. "What need have I to animadvert on so palpable an absurdity," he asks, "as half of 51 laborers to attend on 27 masons?

"What makes it worse," Romans continues, "on this very day on which I write, I am reduced to the dilemme of keeping only 7 masons on the principal work. The other 20 are employed in breaking and carrying stone, by reason that all the labourers are employed in unloading of vessels at the pier head. Where then is the wonder, that we advance not as we should do, and that the expence becomes great?

"We are on an island where we have not a single stick of timber fit to do any thing with, except making firewood, and not even that, as most is a shrubby kind of pine," Romans complains [today Constitution Island is once again heavily wooded]. The carpenters shipped up from New York City, he says, are less diligent than the "country carpenters, who labour harder and do not stand upon the punctilium of stated hours."

The city carpenters, continues Romans, are not familiar with the kind of heavy timber dressing required for his odd-shaped blockhouse. They receive timbers "hewn truer than they are able to do it; then to reduce it to what they judge to be a true square, they line it and hew half way down; afterwards turn it, line it again, and hew the other half way; thus a piece of timber is lined eight times, and hewed to these eight lines, to make it worse than it was.

"Next the piece," one can hear Romans' voice shrilling, "through lack of oxen, is drawn by 20 men, to where it is wanted; with regard to a road, your landing was in a wrong place before I came here. Then the carpenters discover it not to be hewn in the square, and line it and hew it again eight times over; then the dovetail is cut, and when put together, they see it makes bad joints, therefore they have to do it over again. Whereas the country carpenter (used to such work) hews the whole side through by one line, and thereby leaves it, when he parts with it, truer than the others (unacquainted with such work) can possibly do."

To which the Commissioners properly retort: "We should be glad to know who but yourself constructed, ordered, and directed" the building of the blockhouse? "Had the timber for it been ordered in pieces of length equal to the sides, instead of pieces eighteen feet long, unwieldy for men to move and bring up on the rock, we should not have seen so great a waste of timber sawed from every piece, laying about the works; no trifling ordinary expence this."

"I am utterly at a loss," Romans replies, "for what you mean by a waste of timber. My order of 18 feet long was right. The day I got your epistle, I looked around the works for waste timber; I saw none but chips. The truth is, gentlemen, you have no business with my calculations of that kind; you are only to judge afterwards.

"Much against my inclination and advice," Romans continues, "256 iron bolts have been used instead of so many trundles [treenails or trunnels; dry wooden pegs to

join timbers, wet later for tight fit]. These weigh each 5 lbs.,
thus 1,280 lbs.at £28 per ton: £17 18s 4d. Instead of those
256 bolts, as many trundles would not exceed £1 10s. I had
ordered 2,000 of oak. Locust was not to be had." Then, to
twist the knife a bit, "I do not know what ruined your credit.
The badness of that among the country people prevents
our getting anything regularly."

The Commissioners reply: "Your calculation of the extra
iron in lieu of trunnels may be just, but you will be pleased
to observe it was a case of absolute necessity, as you had
never mentioned anything of trunnels until they were
wanted, and then informed us that only locust trunnels
would answer. Those being not to be got after many
applications up here," they continue, "the blockhouse
could not be left waiting for them, as our orders from the
Provincial Congress, dated the 28th of September, was to
get at least twelve guns directly mounted for defence, which
you were frequently urged to perform. We were frequently
expostulating with you on the backwardness of our having
some place of defence finished.

"All we have to say further on this head," the Commissioners conclude, "is that we were of the opinion, and told
you so, that there was no necessity of making a temporary
work have such an elegant outside appearance, and the
inside to be lined with so much nicety and expence."

"What you mean by an elegant outside appearance, I
cannot conceive," Romans snaps back, "except indeed, the
rings and staples outside of the ports; which I never thought
of, nor would have had them there had you not ordered
someone to put them there.

"About the 'inside lined with so much nicety and
expence,' " he adds, "the extra cost of this is no more than a
day and a half's easy work for a carpenter (about 14s.); for it
must be lined, let who will live in it. And as it generally is the
residence of an officer, as it is now for me, I thought that
passing the jack plane over one side of the lining was the
least that could be done.

B. Romans.

Scale of 1500 feet

150 700 300   600   900   1200 1500

Steep and Inaccessible Precipices

Very rough broken ground

Impassable Marshes and a Creek

R I V E R

26b. The same map (engraved for Force 1837).

Scale of 1500 feet.

River:—

Steep & Inaccessible Precipices

Very rough & Broken Ground

Impassable Marshes and a Creek

26a. Romans' sketch map of Martelaer's Rock 1775.

"What would you have said," Romans goes on, "had I lined the roof, divided the lower room with panel partitions, and put up a panel ceiling to remove and put up at pleasure? For most block houses are so, and the meanest are partitioned. It seems your idea of a temporary work is, that it ought to last six months— then to build a new one again.

"Am I the paltry human being that is not allowed to direct his own plan? I cannot omit mentioning," he concludes, "that when I first took a superficial view of the ground, I judged it to be less rough than it proves to be." At that, the Commissioners move in: "The superficial views you say you took of the ground when you first came up here was a great error. It should have been minutely examined, to have made a proper estimate of the works to be erected." And: "In your estimate of the expenses of the barracks, you forgot iron, and glass for the windows."

"You catch at my word 'superficial,' " Romans retorts, "as drowning people do at straws. I will tell you something, perhaps to you extraordinary. What I call a 'superficial' view was such as most other surveyors would call a perfect survey. I am from long experience enabled to take more exact surveys of places with a piece of paper and pencil, than perhaps 99 beside me can with all the circumstantial apparatus generally used. My business, well followed, is three men's work. It is true," Romans confesses, "I forgot the iron for the barracks. Glass is a trifle; 318 panes cost about £6 12s. 6d."

The Commissioners open a new tack: "We wish you would consider how often we have requested you to send from this post your negro, which we now insist on, who is a nuisance, and has caused much dissatisfaction amongst the people." Romans hits his [unpaneled?] ceiling: "You proceed to a matter, which, had I not been convinced of the integrity of your transcriber, I could never thought would have proceeded from you. It looks so much like the little vengeance of disappointed scolds. The negro is more rogue than fool. It is hard indeed that I, who in my private station have for many years past never been without a servant, or even two or three, should be raised to a public station to be debarred that privilege.

"Perhaps you think me your officer," Romans continues, at his best, "Softly, gentlemen; that will never do. The Congress appointed me to a rank I esteem more honourable than any I ever held, yet for 14 years back I have been sometimes employed as a commodore [this was not necessarily a military rank] in the King's service; sometimes at the head of large bodies of men in the woods; and at the worst of times I have been master of a merchantman fitted in a warlike manner.

"The rank I now hold," the Colonel-*manqué* continues, "endangers me of being made the butt against which all resentment may break." Even "a rascal, who does not vouchsafe to lift his hat to us, nor even avoids to insult us, comes into our innermost recess, and interrupts us, perhaps at a time when we are consulting the welfare of the community."

"Sir," the Commissioners smoothly conclude the exchange, "we have no other desire than to treat you as a gentlemen who has an important trust committed with us to your charge; therefore, to avoid every thing that should interrupt the harmony that ought to exist among persons employed in the cause of American liberty, we do desire that when you want to have any piece of work done, we may come together and consult about it, that we may approve or not."

To Romans, this reassertion of the Commissioners' arguable right to review and approve "any piece of work," which they wrapped so carefully in an appeal to his patriotism, was a last straw: "I interrupt none of your powers; I meddle with none. But you have hindered me from having much work done, and until I am sole director of my plan, things cannot go well. None can be more happy in the union that you mention. But if I must be cap in hand, gentlemen, to be an overseer under you, it will not do, depend on it. I have too much blood in me for so mean an action, and you must seek such submissive engineers elsewhere. I am, gentlemen, [&c.] B. ROMANS."

A week later, the battle on Constitution Island swelled to a climax; no one knew better than Romans how far his grandiose plans had fallen behind schedule. The first snowflakes of winter would signal an inglorious end to his career as a revolutionary military engineer. Instead of obstructing the Hudson, Romans and the Commissioners had only succeeded in obstructing each other, and to them he addressed this parting shot:

Martler's Rock [Romans even refused to dignify the Commissioners' christening], 16th Nov. 1775.

"To the Commissioners for the fortifications in the Highlands.

"GENTN.— I forebore to make use of the many polite appellations, such as scoundrel, villain &c., with which Mr. Bedlow was pleased last night, so copiously to honor me in public. B. ROMANS."

A week after Romans had tossed down that final gauntlet, Livingston's Committee to the Northward, en route to Albany and Ticonderoga, rowed eight miles downriver from New Windsor to the fort in the Highlands. On November 23rd they reported with distress to the Continental Congress that although there were now more than 70 cannon (still nothing heavier than a 9-pounder) at Fort Constitution, the installation— with its two-company garrison of 100 militiamen— was "in a less defensible Scituation than we had reason to expect, owing chiefly to an injudicious disposition of the labour, which has hitherto been bestowed on the Barracks, the Blockhouse & the South West curtain. This, Mr. Romans assured us, would be finished in a week, & would mount 14 Cannon, but when Compleated we consider as very insufficient in itself to answer the purpose of defence, tho it is doubtless Necessary

27. Remains of Fort Constitution's 1775 southwest curtain rampart — "Romans' Battery" — from West Point's "Flirtation Walk" [Hudson in foreground] (1983).

to render the whole fortification perfect. But as it is the least useful it should have been the last finished."

The ruin of that southwest emplacement, called "Romans' Battery," still clings to a cliff above the Hudson. The three Congressmen noted how "It does not command the Reach to the Southward, nor can it injure a Vessel turning the West point, & after She has got round, a small breeze or even the tide will enable a ship to pass the Curtain in a few minutes. The principal Strength of the fortress will consist in the South Bastion on which no labour has as yet been bestowed. The block house is finished & has 6 4-pounders mounted in it, & is at present the only Strength of the fortress [after more than two months' work]. The Barracks consist of 14 Rooms, each of which may contain 30 men, but they are not yet Compleated for want of Bricks with which to run up the Chimneys." With freezing blasts beginning to whistle through the Highlands— "high wind & ice made on Oars," noted Committee member Robert Treat Paine— Romans was apparently still having trouble with the cost of bricks.

"The Fortress," the Committee continued, "is unfortunately commanded by all the Grounds about it & is much exposed to an attack by Land, but the most obvious defect is that the Grounds on the West point, behind which an Enemy might land without the least danger, are much higher than the Fortress. It seems necessary that this Place on the opposite Shore should be occupied & batteries thrown up. Mr Romans informs us of a place about 4 miles lower down the River where the elevation is much greater. Had we more time, we should have gone & examined it. We cannot help wishing, when we consider the importance of the object," the Committee ended modestly, "that Congress send Some persons better versed in these matters than we are."

In New York and Philadelphia, everyone exchanged fears of a combined British water-and-land operation up the Hudson. A growing conviction that Romans' works would be powerless to stop the enemy lent extra impetus to considering the alternative site six miles to the south. The Commissioners took a further look at "Pooploop's kill," and wrote from Fort Constitution to the Provincial Congress on December 7th that they "found its situation the best by much for any defensive work in the Highlands. No enemy [they unfortunately thought] can land at Haverstraw and cross the mountains to annoy it by land. The river is not much wider over to Anthony's nose than here [they were right]; a battery of heavy cannon would command it down and up, the length of point blank shot. From Pooploop's kill there is a tolerable road to the West Point; an enemy might bring cannons by land against this post."

The New York Convention finally decided the time had come for a change. On December 14th, ostensibly to "accomodate the Difference subsisting between the Commissioners & the Engineer," the legislators took up the question of whether Romans had "either Mistaken the Charge committed to him, or as appears from his Conduct, has assumed Powers with which he knew he was not intrusted." Returning from a survey of the works on Constitution Island, a new three-man committee consisting of Col. Francis Nicoll, Col. Joseph Drake and Thomas Palmer (a civilian with a reasonable amount of engineering skill, later placed in charge of constructing the works above Popolopen Creek) submitted an eight-page report to the Convention.

Palmer documented every possible way in which the design of Fort Constitution appeared inadequate, and urgently recommended establishment of an additional barbette battery of eight 18-pound cannon, without fancy embrasures, on the gravel hill near the eastern end of the island, to "sweep the river southward." It was something Romans had actually considered (on two of his sketch maps— *Figure 24 "H,"* and *Figure 23 "B"*) and then proceeded to ignore.

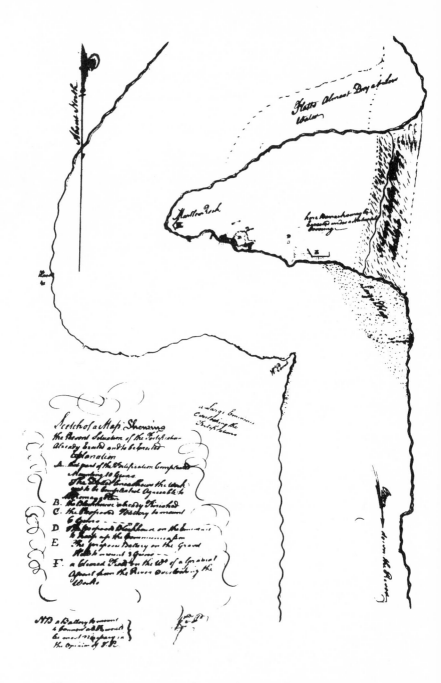

28a. Thomas Palmer's proposed strengthening of Fort Constitution 1775.

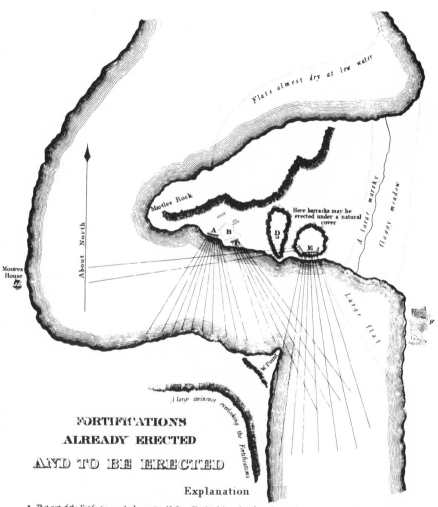

28b. The same drawing (engraved for Force 1837).

After lengthy discussion, the Convention finally crossed the Rubicon. They decided "Mr. Romans was to blame in refusing to consult the Commissioners on every Matter of Importance, before he attempted to carry it into execution." Palmer was immediately dispatched (together with Fortifications Commissioner Grenell) to woo the Continental Congress in Philadelphia with sketches of his own projected improvements to Fort Constitution.

Palmer was also requested to show how Romans had completely ignored "two large Eminences overlooking the works, so situated that an Enemy might improve them much to our damage"; and that when the committee noted to Romans that "it was his indispensable duty accurately to have observed those matters in his first report to the Continental and Provincial Congresses, he answered that he had pointed out the necessity of the one, & the other he had but lately thought of." Sufficiently emboldened by news of Palmer's mission to start overriding their "master workman," the remaining Commissioners at Fort Constitution began to divert available construction materials downriver.

Romans' time was running out.

Two days later, British naval intelligence finally caught up with events along the Hudson, among other items carefully noting— as Romans had long feared— that his name still remained on the royal payroll. Navy Captain George Vandeput supplied Captain Hyde Parker, Jr. in New York City with the belated and disquieting information that "a Fort is built about sixty miles up Hudson's River, intended to prevent the communication with Albany." Parker little realized the installation's ineffective condition as he conveyed Vandeput's news to Vice-admiral Samuel Graves in Boston, also noting, "Bernard Romans is appointed Engineer of the fortress in the High lands, he had 50 Pounds a Year from the Crown as Botanist for One of the Florida's." The British bureaucratic net began to close on Bernard Romans' £50 "pension."

On December 18th Governor Tryon wrote to Major General William Howe in Boston: "Bernard Romans is appointed engineer of a fortress on the Hudson River." On January 16th Howe forwarded Tryon's letter to Whitehall (where two months before the intransigent Lord George Germain had finally replaced the vacillating Dartmouth as Secretary of State for the American Department). And by early February, the Earl of Sandwich at the Admiralty received the identical news from Vice-admiral Graves and passed it on to Germain: "Bernard Romans appointed engineer in Highlands."

On February 27th, His Majesty's Commissioners for Trade and Plantations, meeting with Germain, "read a memorial of the agent for West Florida [John Ellis], stating he had intelligence that Mr Romans, for whom provision is made upon estimate by Parliament of fifty pounds per annum, had joined the rebels." The agent "prayed directions of the Board, whether he should issue payment on a bill of exchange drawn on him by the said Mr. Romans?" (Romans had apparently moved quickly to insure some final payment).

The Lords of Trade directed that "in consideration of the intelligence above mentioned, the agent would be fully justified in with-holding payment of the said bill." The word was out; it even reached as far as Pensacola. The engineer's old patron Governor Peter Chester belatedly warned Whitehall in April: "Bernard Romans is believed to be in the service of the rebels."

Romans' 1775-1776 payment was removed from the West Florida spring Estimate. Not until August 7th, however, did Germain finally reply to Chester: "I am sorry that Mr Romans has made so ill a return to the kindness shown him by Government. As an unfaithful subject of His Majesty, he no longer deserves countenance or protection. Allowance to Mr Romans is discontinued."

29. Site tablet marking Capt. William Smith's 1776
"gravel hill" battery, looking downriver (1983).

CHAPTER VI

Congress Intervenes

*In which a disillusioned Congress removes
Bernard Romans as engineer of the failed fort in
the Highlands, he is assuaged with command of a
company of Pennsylvania artillerists, and the seat
of war finally moves from Boston to New York.*

Three days into the new year of 1776, Bernard Romans
was summoned through the snows of Westchester County
to a meeting with the New York Committee of Safety—
again "appointed by the Provincial Congress to act in their
recess." Pierre van Cortlandt, the new committee chairman,
was obviously eager to rid the Province of a troublesome
antagonist. He ingenuously converted Romans into a
messenger to the same Philadelphia legislators from whom
he had come five months earlier. Van Cortlandt asked him
to bear a group of "Drafts, with necessary Information" on
the current state of his Hudson River fortifications.

It was a tense moment for Romans; he may not even have
known the contents of the Committee's letter he carried to
"The Honble. the President of the Continental Congress"
John Hancock. Couched as an appeal to America's purse
strings, the letter purported to convey Romans' own "Ideas
of the Matter" that, far from being even partially successful,
Fort Constitution "cannot be rendered sufficiently Secure
for a Lodgment of Troops, and to answer the End of a
Fortified pass, without more Expence than our Commissioners appointed to superintend that Business think
prudent. Besides which they observe, that his Scheme
cannot be compleatly executed with that dispatch the
Service may require."

Van Cortlandt's letter also noted that New York's earlier
messenger Thomas Palmer, now a rival to Romans, "has, to

remove the Difficulties occasioned by a Difference of
Opinion between the Commissioners and Engineer, doubtless pointed out to Congress certain places on the River
which would better answer the purpose of a Temporary
Defence, and at much less expence than will necessarily
attend the Execution of Mr. Romans' Scheme."

Van Cortlandt then tossed Romans' uncertain future
squarely into Philadelphia's lap: "We submit it to the
Consideration of Congress how far the Completion of the
Fortifications already begun ought to be carried on upon
Mr. Romans' plan; or that which has doubtless been
proposed by Mr. Palmer as the sense of our Committee who
were sent to view the works." Butter melts in the
Chairman's mouth as he continues: "We hope that this
application, while Mr. Palmer is attending on the Congress,
will not be construed to his disadvantage. Had we been
possessed of Mr. Romans' plan in its present compleat
State, and had he been in Town at the Time of Mr. Palmer's
Departure, we should have required his accompanying that
Gentleman.

"As there is some prospect that Mr. Romans may reach
Philadelphia before the plan of Fortification is finally
determined by Congress," van Cortlandt innocently concludes, "we should think ourselves inexcusable for with-
holding from them any Means for enabling them to
determine so important a Matter, on the best Lights in our
power to furnish."

As van Cortlandt hoped, Congress had begun to debate
the real usefulness of Fort Constitution while Romans was
still en route to Philadelphia. It took up the "Report of the
State of New York" and Palmer's own plan for either
redeeming the situation— or abandoning the fort entirely
in favor of new works at at Popolopen Creek. The diary of
Richard Smith, delegate from New Jersey, records: "Great
Fault found with the Fort now constructing in the
Highlands under the Auspices of Bernard Romans the
Engineer, as too large and expensive and ill calculated to

annoy the Enemy— Maps produced and Proceedings of N
York Convention thereon read. Mr. Palmer attending on
their behalf, ordered that he be heard Tomorrow Morning."
The legislators then followed up with a discussion of
possible alternatives.

On January 5th, Congress finally turned a corner. It
"Resolved: That for the reasons contained in the papers
hereunto annexed, no further Fortifications be erected on
Marteler's Rock on Hudson's river, and that a point of land
at Puplopen's kill, on the said river, ought without delay to
be effectively fortified." Smith recorded in his diary how
Thomas Palmer and "Capt. [John Grenell] from New
York" had been "called in & examined. This affair took up
several Hours. Was at last resolved to abandon the Works
at Martilers Rock, with Romans the Projector of them."

By then Romans himself had arrived, carrying the
Committee of Safety's letter. As van Cortlandt suspected,
the engineer found his reception in Philadelphia very
different from that of a few months earlier. The extem-
poraneous nature of the war was changing; Congress had
established a reasonably well-organized pattern of army
command, and early volunteers like Romans, who had been
found wanting, were rapidly being replaced.

To Romans' embarrassment, he discovered he had
become something of a political pariah with questionable
fame now denounced in both civilian and military circles.
He was kept waiting eight days while a special Congressional
committee debated (according to Smith) "whether Bernard
Romans shall be called in and examined about his
Fortifications on Hudson's River." The committee finally
decided not to bother, and ordered Oliver Wolcott,
delegate from Romans' adopted state of Connecticut, to
meet and tell him that his services as a patriot engineer were
no longer required.

What had begun as the grandest period of Romans' life
was now drawing to a bitter end. Still, he was not without
Congressional support. Amidst all the swirling acrimony,

John Hancock apparently felt some residual loyalty to the man who, among other things, had graphically immortalized his Boston mansion. Even Robert Livingston, whose Committee to the Northward had been so critical of Romans' work at Fort Constitution, was having second thoughts.

"I must confess," Livingston wrote to Thomas Lynch, the South Carolina Congressman, "that desisting from the work at Martilars Rock after so much money had been layd out upon it surprized me. The finishing only one bastion would make it extreamly formidable and indeed impassible unless the enemy were very much favored by the wind, nor could the expence be very great after what has been already done. At least let a battery be erected at some little distance from the fort to the eastward, which may be made of earth by the soldiers, & commands the whole reach below & must of necessity prove very troublesome to shiping coming up. I wish some Engineer could be found to take charge of this matter at any price." But such an engineer could no longer be Bernard Romans.

In New York on January 16th, the Committee of Safety ordered all useable materials still at "Martlaer's Rock removed at the first opportunity to Pooplopen's Kill." Romans' pretentious military complex had begun to die.

In February, Congress finally took up the "petition from Mr. Bernard Romans," but only to send the engineer's problem back north; it advised the New York Convention "to pay Mr. Bernard Romans up to the 9th day of this month." Then, on February 15th, troubled by Livingston's misgivings, Congress changed its mind once more about Martelaer's Rock and decided to build Palmer's recommended barbette after all. Captain William Smith, an experienced engineering officer under General Charles Lee in New York City, was assigned to the new construction. Smith immediately laid out lines for a raised semicircular earthen gun platform on the identical spot Romans had once indicated— and then ignored.

*   *   *

Brigadier General Horatio Gates, Washington's querulous adjutant at Cambridge but a professional military man for more than a quarter of a century, noted the obvious: "As soon as our artillery makes the town of Boston uneasie to its present garrison, so surely will they leave it. Depend on it, the ministerial push will then be to regain possession of the rivers St. Lawrence and Hudson." Gates hoped aloud that there would be "a good fort somewhere in the narrows above the Highlands to command the channel."

During the winter, as had been planned, Henry Knox— the Boston merchant/bookseller become artillery colonel in the Continental army— succeeded in loading 60 tons of the cannon Romans helped inventory at Ticonderoga onto "slays." With 80 yoke of oxen, Knox hauled the precious armory more than 300 miles— down the Hudson Valley to Claverack, then eastward across Massachusetts snows and frozen rivers to the south Boston suburb of Dorchester.

Quickly installed on Dorchester Heights (behind straw-filled "chandeliers" ingeniously devised by Colonel Rufus Putnam), the guns effectively checkmated the British forces and ships in Boston and the harbor below. Meeting in a council of war with his general officers at Roxbury MA on March 13th, Washington said he had "reason to believe the Ministerial Fleet and Army" would soon evacuate Boston, "destined for New-York, and would attempt to possess themselves of that City, command the Navigation of Hudson's River, open a Communication with Canada and cut off all intercourse between the Southern and Northern colonies."

Four days later, General William Howe did sail out of Boston, but headed north for Halifax to regroup with his older brother, Viscount Richard Howe, commander-in-chief of the British Navy's North American squadron, for more vigorous prosecution of the war. Patriot military attention immediately shifted south to New York; Washington arrived in that city on April 13th.

Ten days before, Catholic Charles Carroll of Carrollton, with aging and ailing Benjamin Franklin, set off from Philadelphia on a fruitless propaganda mission. They were attempting to salvage through personal appeal to the independent nationalism of Canadian Catholics, what had become a patriot military debacle along the St. Lawrence River. The Carroll/Franklin party sailed past "Constitution fort" on a wretched trip up the Hudson River in sail-shredding early spring winds.

They went ashore "from curiosity" on April 5th to find a much-revised, simplified Fort Constitution, now down to only 22 cannon— the remainder had been dragged south to Popolopen Kill. They also found "not one Gunner or Artilleryman" to serve them, and "only 102 men fit to do duty." There were "81 Quarter Barrells and 1 whole Barrell of Powder" in spring-soggy magazines that were proving dangerous to dry out— even with controlled fires. The soldiers were demanding hazard pay.

It was a foretaste of the problems that, for the next 18 months, would plague the unhappy island and its discouraged, underfed, ill-equipped, usually unpaid and drunken garrison. (Mainly overlooked, due to everyone's miserable living conditions, was the December order from the Convention prohibiting "any Sutler or Retailer of Liquor of any kind to remain at or near the said Fortification, amongst the Mechanicks, Labourers, or other persons.") Before long the original garrison of local militiamen rebelled; it was replaced with "more dependable" Continental regulars.

Carroll was highly critical of Romans' construction plans and actual work: "It does not deserve the name of a fort, being quite open on the north-east side. If Lord Howe knew its weak state," he wrote by express to Washington's headquarters in New York, "he might take the fort in its present situation with 60 men and without cannon. He might land his party a little below the fort on the east side, march over a marsh, and attack the back" (much as the patriots would do themselves three years later at Stony Point).

Carroll also noted how the additional barbette battery ordered two months earlier by the Continental Congress had been "strangely neglected"; only a portion of the gravel hill had been leveled. Understandably, military interest was shifting "lower down the river," Carroll reported, "to a battery to be called Fort Montgomery, and another a little below Cape St. Anthony's nose [Fort Clinton]."

With a huge British fleet expected momentarily in New York harbor, Washington took an urgent interest in the Hudson River fortifications. Responding to Carroll's report and seeking first-hand opinion from experienced military personnel, the Commander-in-Chief ordered an investigation by a committee that consisted of stodgy but dependable Brigadier General Lord Stirling, Colonel Rufus Putnam, now Washington's most dependable engineer (he had trod in Romans' footsteps before— as deputy surveyor for West Florida in 1774) and Captain Winthrop Sargent, a skilled artillery officer. "I have great reason," Washington wrote to Stirling, "to think that the Fortifications in the Highlands are in a bad situation; and the garrison worse."

The group's report from Fort Constitution underscored all previous opinion, and forever branded Bernard Romans' military engineering work as inadequate. "The westernmost battery," asserted Stirling, "is a straight line constructed by Mr. Romans at a very great expence; it has 15 embrasures which face the river at a right angle, and can only annoy a ship going past." Although other parts of the completed fortification, Stirling continued, "look very neat and picturesque, upon the whole Mr. Romans has displayed his genius to very little publick advantage. The works in their present open condition and scattered situation are defenceless. Everything on the island is commanded by the hill on the West Point."

30. Remains of Fort Constitution's 1776 southwest curtain rampart —
"Romans' Battery" — from: (a) a beach on Constitution Island
[West Point at right]; (b) within the lines of Constitution Island
[West Point in background] (1983).

Stirling was evenhandedly critical of the manner in which the Commissioners for Fortifications were discharging their own responsibilities (at "eight hundred dollars a month," he underlined to Washington). "One good engineer, with artificers from the Army," Stirling suggested to the commander-in-chief, "might, I think, do the whole business as well."

The day before Stirling made his report, Samuel Brewster, chairman of the Safety Committee at New Windsor, six miles upriver, bravely brought a different kind of complaint to the New York Convention. It concerned Convention member and Fortifications Commissioner Jonathan Lawrence, as well as Mrs. Lawrence and their son Jonathan Jr., whom his father had made "Clerk of the Check" at Fort Constitution. "Mrs. Lawrence early in the spring opened a shop here," Brewster alleged, "Then, knowingly contrary to the resolve of Congress, she retailed Bohea tea at eight shillings per pound; and to evade the resolve, pretends to sell her tea at six shillings per pound, but will not let the purchaser have the tea unless he takes a paper bag to put it in at two shillings.

"Upon having this complaint laid before us," Brewster told the Convention, "the remainder of the tea was removed to Fort Constitution under the care of her husband and son, which has much disgusted the inhabitants, as they conceive it as a kind of an insult to have that fortress made an asylum for that useless herb and the illegal venders of it."

That incident probably triggered Washington's desire for a change. He requested a committee from the Convention to wait upon him, and when they returned, it was with the information that the Commander-in-Chief "will not have occasion for the further service of the Commissioners, as he will take the sole direction of those Fortifications." Whereupon the Convention eventually "Resolved, therefore, That the Commissioners employed to superintend the

Fortifications in the Highlands be, and they are hereby, discharged from that service."

Civilian control of the situation had been superseded.

*   *   *

Languishing in Philadelphia, Bernard Romans attempted to promote any kind of army assignment through whatever friends he still had in Congress. Battered but resilient, he put his delay to typical good use, reconnoitering the ground for a *Chorographical Map of the Country Round Philadelphia* (the map would finally come off press in 1778.)

On February 8, 1776, the Provincial Council of Pennsylvania, "Agreeable to a Resolve of Congress" and probably under gentle pressure from President John Hancock (who may have felt the aging engineer deserved a bit better than he had received), appointed Romans with his original Connecticut rank— captain— to a "Company of Matrosses" raised the previous week to support the faltering invasion of Canada. Matrosses were artillery soldiers serving as assistants to the more-skilled men who trained and fired the cannon; they did the dirty work of hauling, sponging and loading.

Fresh commission in hand, Romans appointed his new Captain-Lieutenant Gibbs Jones to organize the company, while he returned north to wind up his affairs. Carrying a private and philosophic letter about the conduct of the war from John Hancock to Governor John Trumbull in Hartford, Romans paused long enough in New York City for a parting shot at the Committee of Safety, who still delayed his final salary payment (through February 9th) ordered by the Continental Congress. "The time is now expired," Romans protested, "in which your humble Petitioner was to have appeared at the head of his company, and want of money prevents. Your humble Petitioner prays an order may be granted him, pursuant to the said resolves of Congress, that he may be enabled to proceed, and save his honour."

It was a strategic error. The Committee turned the letter over to the reconvened Convention, who ordered any consideration "deferred until B. Romans is called in, and interrogated as to the reasons of his so long absence from the Fortifications at different times."

Two weeks later on March 18th, "The petition of Mr. Bernard Romans, attending at the door, was read and filed, and he was called in. He assigns such reasons as he could give for his absence, and for the charges in the account he had made against this colony. After he was heard," the minutes tersely conclude, "he withdrew." New York refused to pay Romans a penny of additional salary. By April 1st the Continental Congress was forced to direct its Commissary to "advance Captain Romans one months' pay, and also subsistence money to Albany."

you against me. the facts as here related are upon
my honor true, ▬ Give me Leave to assure your
honor that I would sooner be guilty of a crime
of a Black dye than to be at the head of a gang
of ruffian marauders — & that I will break
my Sword & give up my Commission that instant
it is out of my power to carry Command & to main,
tain discipline; the people may have been illtrea
ted by the first Comers, But if they carry Suspicion
against all in the manner they did against us it
will be impossible at Some times to prevent ill
Blood in the troops ——— I have the Satisfac
,tion to know that the good order of my Company
had every where else high encomiums bestowed
on it, — & your honor's Secretary can inform
you how he has Seen me handle one for being re.
fractory,

  Excuse this taking up So much of your
time & permit me to assure your honor of
My Strictest regard & most humble respect—
   I am
     Honored Sir   Your honor's most
             obedient & very
             humble Servant

Ticonderoga Landing
the 25th april 1776

31. Signature page, Romans' April 1776 letter to General Schuyler.

CHAPTER VII

Captain Romans Redux

*In which Bernard Romans carries his new
command into further acrimony, serves through the
Saratoga campaign while drawing fresh maps,
resigns from the army to marry and write a new
book. . . and then — mysteriously — disappears.*

Bernard Romans' new artillery command in the Canadian
Department of the war would bring him little additional
fame or satisfaction.

Organization of his company was completed by March
25th and Romans' troops set out from Philadelphia for
Canada two weeks later. Within a fortnight, the Captain was
in difficulty with his commanding officer, General Philip
Schuyler (who carefully avoided meeting Romans *en route*).
Romans faced an Albany court martial on charges that his
troops killed poultry and abused civilians and army
waggoners on their march north from Albany. Specifically
relating the "string of complaints" lodged against Romans,
Schuyler remarked in a letter to Washington on "the
disgraceful conduct of the Army, where officers connive at
the depredations of the men."

"Mr. Romans' conduct, by all accounts, has been very
extraordinary," Benedict Arnold reported in from Canada.
By virtue of valorous military exploits Romans must have
envied, Arnold was now a brigadier general, on his way to
becoming one of the Continental Army's most talented
field commanders.

Romans' full letter of explanation to Schuyler from
"Ticonderoga Landing, the 25th april 1776," is an archetype of
his aggressive defensive style:

"I am very sorry that I missed the honor of an Interview
with you to deliver the Inclosed," wrote Romans to his

commander, "which I promised Mr. Hancock that I would do *propria persona.*"

(Hancock had written Schuyler: "I take the Freedom to mention to You Capt. Romans the Bearer of this. He is appointed Capt. of an Artillery Company but in the Opinion of Many his Abilities are not fully Call'd forth. He is said to be well skilled in the Engineering Branch, of which on Conversation You will be the best Judge, & I dare say, from my Knowledge of Your Disposition, so far as You find him Capable, that You will be ready to recommend him & promote him." Hancock added, "You will excuse this Mention," unnecessarily, perhaps, because Schuyler had no desire to see Romans in any case.)

"I arrived here Last Night," Romans' apologia to Schuyler continued, "being the 17th day since we Left the City of philadelphia, with a company of Artillery under my command; every where acknowledged to be a fine set of men, and I have the pleasure to say, already tolerably expert in the use of Artillery. We are not fully supplied with arms; Mr Hancock told me that from you I was to expect the remainder— this is the only difficulty I now labor under.

"Urged by the strict injunction I received to expedite my march, I have made it in the short time mentioned. I humbly apprehend that I need not wait farther orders to proceed to Quebec— In the meantime if Your Honor would be pleased to write a word or two in my favor to the Commanding officer [General David Wooster] with respect to my being employed as Engineer [Romans was never one to abandon hope], it will do me an immense favor. I have a right to expect this from Mr Hancock's promises, and have therefore taken pains to have a Captn. Lieutenant [Gibbs Jones], who is well experienced and much used to command— But all this I humbly submit to your pleasure.

"I am now to enter on a Subject very disagreeable to me. I appeal to every housekeeper who entertained us on our march for a character of our people, yet the people on this side of Albany have been the cause of vast trouble to me, nor have I lost a man on the whole route until I came on this side of that city.

"In revenge for my preventing the tippling houses to sell their infernal rum to my people [an interesting adjective from a man who two years earlier had been willing to pledge "every my Effects especially the Copperplates of my work" for "two Hogsheads of Rum by me received"], they encouraged them to desert, whereby I lost four of my men. The day before yesterday, as we came along by one Graeme or Graham, I was suddenly surprized by the officer in the Center calling aloud that the rear was in disorder; this being the first instance of our line being broke on the march, I flew to the spot. When I had appeased the ferment, I enquired the cause & found that some of the men had asked for a drink of water & were refused.

"The consequence was a general cry of A Tory! A Tory!, & had we not had the strict command we carry, I hardly believe the fellows house would have stood five minutes. Among other mischief done, a turkey was killed. I offered every satisfaction in my power, & Mr. Jones my Captn. Lieutenant spent near half an hour to persuade Graham to take pecuniary satisfaction for what he suffered. But all to no purpose; he said he would complain & I should pay dear for it.

"Again at Wing's, while we took a morsel of dinner, a complaint came that our men had killed his fowls. I flew up in a rage, & called the men together, and found none missing. I sent the third Lieutenant and two serjeants to see what was the matter, & found that the whole had arisen from two men withdrawing behind the barn, & they had been mistrusted by some people who were placed to watch them. At length the whole family acknowledged that not a fowl had been touched—

"The affair as I thought was settled, but I was scarcely seated before I heard another alarm, & was told one of the men had struck the Landlord. The fellow excused himself by saying the family had joined abuse of the officers to a

false accusation. I then gave him a most severe bastinado, &
offered further satisfaction. But to no purpose; he again
would complain— and I suppose both these complaints
have now reached your honor.

"I dwelt this long upon this disagreeable subject, that you
might be timely informed of the circumstances & that
misrepresentation might not prejudice you against me. The
facts as here related are upon my honor true. Give me Leave
to assure your honor that I would sooner be guilty of a
crime of a Black dye than to be at the head of a gang of
ruffian marauders— & that I will break my Sword & give up
my commission that instant it is out of my power to carry
command & to maintain discipline; the people may have
been illtreated by the first comers, But if they carry
suspicion against all in the manner they did against us, it will
be impossible at some times to prevent ill Blood in the
troops.

"I have the satisfaction to know that the good order of my
company had every where else high encomiums bestowed
on it, & your honor's Secretary can inform you how he has
seen me handle one for being refractory.

"Excuse this taking up so much of your time & permit me
to assure your honor of my strictest regard & most humble
respect. I am, honored sir, Your honor's most obedient &
very humble servant, *B. Romans.*" It was one of the Captain's
most classic epistolary performances.

Eventually the charges against Romans were dropped.
On July 2nd his company was reassigned south from the
Canadian to the Northern Department (bounded by the St.
Lawrence River and a line drawn east along the New York-
Pennsylvania border into New England). Romans was
stationed at Ticonderoga, commanding four guns in "the
old French lines," and continued there despite further
acrimonious disputes with his junior officers.

During the summer, one of them took the liberty of
writing directly to Horatio Gates, the commanding general
who had replaced Schuyler, saying that Romans had

"neither honour, honesty nor valour in him." Perhaps the letter merely called Romans to Gates' further attention. The two had similar prickly personalities (Gates was always a bitter thorn in Washington's side), and both men possessed the same limited ability to command the minds and hearts of their fellows.

In the midst of the acrimony, Romans— that "ingenious man"— was able to persuade the Pennsylvania Committee of Safety to furnish him with "such Materials as he might require to perform an Experiment, in order to give a specimen of his skill in destroying distant Objects by Fire."[!] Romans' customarily acerb correspondence slowly tapered off. A pleasant change in October was his certificate to Congress endorsing special pay for two scouts assigned by General John Sullivan to reconnoiter enemy strength at Trois Rivieres. They "went so near," Romans commented admiringly, "as to have fourteen cannon shots fired at them."

While decisions in the New York Convention and the Continental Congress had destroyed Romans' military reputation as an engineer, London never had difficulty recognizing his cartographic skills. On October 15th— two weeks after British Undersecretary of State William Knox again found it necesary to remind John Ellis that "Mr Romans has joined the rebels and is not to be paid allowance"— Sayer & Bennett, Fleet Street engravers and semi-official printer of British military maps, issued a cavalry officer's "Holster atlas"— the *American Military Pocket Atlas: Such Maps as an Officer may take with him into the Field, especially those of Colonies which now are, or probably may be, the Theatre of War.*

No. 5 of that six-map package was a neatly-folded *General Map of the Southern British Colonies in America,* based in large part on materials Romans had forwarded in prior years to the Colonial Office. "Engineer ROMANS' work," said the map publishers' introduction, "will be found to have a very particular Degree of Geographical and Topographical

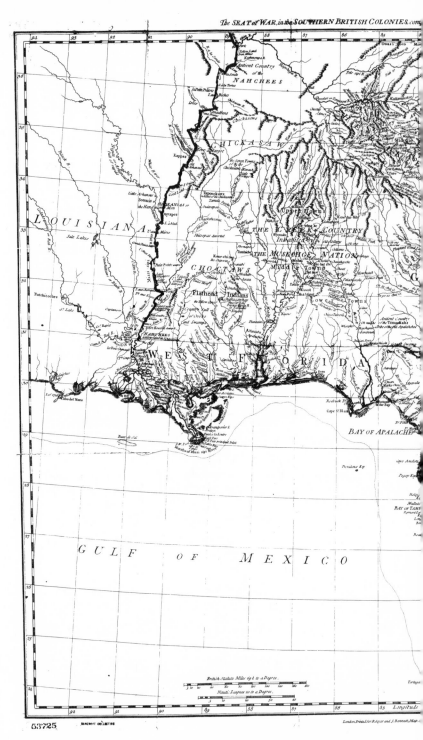

32. *General map of the Southern British Colonies:*

Romans et al. (engraved by Sayer & Bennett 1775).

The inscription text visible on the map (partially legible, handwritten):

The Townships on the East of Lake CHAMPLAIN are those granted by the State of NEW HAMPSHIRE. Those that are marked Y. which were Granted by the State of NEW YORK on such and Grants where they do not interfere with the Hampshire Grants. the Genuine New York grants that interfere with the Older ones are marked with dotted Lines. and as they are mostly granted to Officers in the Regular army except a few which have the names of Wallis. Kemp. and some such other favourites of those Princes of Land jobbers Moore. Dunmore. Colden. and Tryon. Stamped on them, it was not thought worth while to note them. Especially as the Inhabitants of this State of Vermont now hold them by the triple title of honest purchase of Industry in Settling, and now lately that of Conquest.

(North)

33. Romans' unsigned 3½ sq. ft. *Chorographical Map of the*

Northern Department of North America 1778 ["North" added].

Merit." The humbled designer of Fort Constitution thus
became the only American during the Revolution to have
his maps published and used by the enemy.

Late in 1776, Romans' persistent efforts to be reclassified
as an army engineer were finally rewarded by Gates (a
general often "subject to the fascinating and corrupting
influence of flattery" said a close aide half a century later).
The general assigned Romans to "inspect the works" at
Fort Anne [only a dozen miles from his comic victory at
Fort George], as well as those at nearby Skenesborough.

Romans' on-the-job training at Fort Constitution had not
been wasted. At Skenesborough, the Captain reported he
found a "thing called a fort, baffling all description; an
'irregular polygon,' irregular indeed, and by its form,
indefensible with a vengeance." His critical comments to
Gates, transmitted "with all possible esteem" and accompanied by detailed sketches— the kind of drafting Romans
did so well— actually contained valuable defensive sug-
gestions that foreshadowed some of the tactics effectively
used the following year to hamper Burgoyne's advance.

On the first day of the "Year of the Hangman"— whose
last three "7's" offered a solemn portent of the royal gibbets
awaiting unsuccessful patriots— Gates authorized detached
service for Romans, but with no increase in rank; the
matross company was permanently turned over to his
Captain-Lieutenant. In his usual fashion of taking carto-
graphical advantage of a region in which he found himself,
Romans now began to compile— for "real Lovers and
Encouragers of the Geographical Science"— a complicated
four-square-foot *Chorographical Map of the Northern Department of North America / Drawn from the Latest and most accurate
Observations.*

(As the war dragged on, Romans' maps continued to be
pulled from the engraver's press. In June 1777 Nathaniel
Townsend published in Norwich for 8s. [or $2] Romans'
handsome four-square-foot *Map of the State of Connecticut,
with the parts of New-York, New-Jersey and Islands Adjacent, that*

34. 1780 Dutch copy of Romans' *Connecticut and parts adjacent 1777*.

has been the Seat of War for some time past. As with most of
Romans' maps, subscribers received a slight pre-publication
discount. Wholesale merchandising was introduced; seven
maps were offered to booksellers for the price of six. Hand-
colored versions were extra.)

Romans' new draft covered more than 60,000 square
miles with reasonable precision— "from Red Hook, on the
North-River, (the place where the late map of Connecticut
left off) to Three-Rivers, in Canada, and from the Heads of
Merrimack River, in Hampshire, to the Heads of Delaware
and Susquehannah, Westward, including Fort-Schuyler,
and the Oneyda Lake, with Part of Ontario." The map's
patchwork appearance delineates the boundaries of several
hundred New York and New England townships, large
private land grants and manorial landholdings "taken from
real surveys, performed by the Publisher himself, com-
municated by his friends, or collected from the Secretary's
office in New-York [City or State Romans does not specify;
the capital was now at Kingston]."

"In every part where those fail," Romans' ad in a Boston
newspaper reassured potential subscribers, "the work is
founded on the best maps, corrected by the publisher, from
observations and remarks made during his frequent travels
in every part of America." The new name "Vermont" had
been used for the first time in April 1777 in a broadside by
Thomas Young of Pennsylvania, addressed to the inhabitants of the Hampshire Grants; Romans' Northern Department map is generally accepted as its first cartographic
application.

The map's orientation is uncommon— for Romans or
any other mapmaker. For reasons not clear, he has rotated
North 120° counterclockwise; the country appears as a
bird's-eye view from Montreal towards Boston. In an inset
in the upper left corner he skewers— in typical Romans'
fashion— corrupt New York "Harpy Land Jobbers" for
awarding large grants of Vermont land to their friends,
where previous grants already existed. Below the inset, a

deliberately wide six-line description occupies the area of the still unexplored Adirondack Mountains, "their tops always White with Snow," which Romans predicted "will in future furnish a comfortable retreat for many Industrious Families." (The Adirondack region actually remained poorly surveyed for another century— until 1872.)

The map was first advertised in the *Boston Gazette, and County Journal* as early as November 1777, and the following year in the *New-London Gazette* and *Connecticut Journal.*

By now, Bernard Romans, still tough but aging, had served in the field for more than two years, under conditions that sometimes saw up to a quarter of the Continental Army listed on the monthly returns under the heading "Hospital," and also at a time when only 2% of the civilian population lived to see 65. By any evaluation, Romans had done his share for the American Revolution. He resigned from the Army June 1, 1778, to return to the relative peace and quiet of Wethersfield. There he married 19-year-old Elizabeth Whiting, whom he had met three years earlier. Within nine months, they were the parents of a boy [Romans' second son?] they named Hubertus.

The indefatigable author-cartographer finally returned to the long-promised book reflecting his Dutch heritage. Working from a shelf of historical texts, he completed for Hartford publishers Watson and Goodwin the first of two volumes entitled *Annals of the Troubles in the Netherlands; from the Accession of Charles V, Emperor of Germany.* It was a worthy predecessor to Motley's classic *Rise of the Dutch Republic* to come three-quarters of a century later.

The *Annals*— at $4 a copy— recounted the "amazing success of a vassal people who extricated themselves from the tyranny of the most exorbitant power in Europe." The work, Romans announced, was "A proper and seasonable mirror for the present Americans, collected and translated from the most approved historians in the native tongue." The first volume sported a handsome two-color title page, and was dedicated ("by Permission") to Connecticut

Governor Jonathan Trumbull. It made the most of striking parallels Romans was able to develop between the Dutch and American wars for independence.

Romans' work was thorough. A 120-page Introduction carries the reader from Roman times to the accession of Charles V of Spain in 1506. The remainder of the volume [later claimed by Governor Trumbull to be the first book of more than 100 pages published in Connecticut] covers a century of bloody struggles for the independence of the Low Countries. It also afforded Romans ample opportunity to vent some traditional American anti-papist feelings. The second volume— now one of the rarest of all early American books— continues Dutch history through the death of Admiral Piet Hein in 1629 (that book was not set in type until Romans left Wethersfield for the last time in 1780, and was not published until 1782).

Atypical modesty creeps into the author's preface to the *Annals*: "I HAVE not vanity enough to rank myself in the eminent dignity of an HISTORIAN," writes Romans, "No; Candid Reader, I only beg of you to regard me as a diligent Compiler, and faithful Translator. As a foreigner, it cannot be expected that I should excel in elegance of composition, or correctness of language; especially in a tongue whose idiom, orthography, connexion and pronunciation are, of all others, the most difficult and uncouth to the ear and powers of articulation in strangers." That comment must have amused anyone who had ever received one of the author's elegantly-phrased letters.

Romans, now close to 60 years old, offered a prefatorial benediction on his adopted country: "MAY Heaven smile on your virtuous Struggles, and give you a lasting and glorious name among the nations."

Two Romans maps were finally engraved and published in 1778: the much-heralded map of the Northern Department— for some reason unsigned, and plagiarized two years later by Covens and Mortier in Holland (as was his *Map of the State of Connecticut*)— plus the small foot-square

A N N A L S

O F T H E

T R O U B L E S

I N T H E

N E T H E R L A N D S.

F R O M T H E

A C C E S S I O N OF C H A R L E S V.
E M P E R O R O F G E R M A N Y,

I N F O U R P A R T S,

A proper and feafonable MIRROR for the prefent AMERICANS.

COLLECTED and TRANSLATED from the moft APPROVED HIS-
TORIANS in the NATIVE TONGUE.

BY B E R N A R D R O M A N S,

V O L U M E I.

H A R T F O R D:

Printed by WATSON and GOODWIN, for the AUTHOR

M.DCC.LXXVIII.

35. Watson and Goodwin's title page for
Annals of the Troubles, etc. 1778.

Chorographical Map of the Country Round Philadelphia noted in Chapter VI, which was also the first American map ever to identify the 1777-8 "Valley Forge Grand American Winter Camp."

A January 1778 advertisement by Romans in the *Connecticut Journal* offering his Northern Department map for sale crankily rebuked "those who have only wrote their names, with a view to shine in a Subscription Paper. They may save themselves the trouble [of calling for their maps], as there are none printed for such."

A letter from Romans to a local merchant, carried from Glastonbury (across the river from Wethersfield) in January 1779, was accompanied by "three or four dozen" copies of his newly-published *Annals* on wholesale consignment— 8% off. The letter enclosed $100 from Romans in payment for a $120 brass kettle (the Continental dollar was sliding rapidly) ordered from the merchant's brother. The author explained he was "somewhat scant & being newly settled, have a good many drains on me." But with customary optimism Romans expressed "no doubt that the books will furnish the other 20 dollars in a day or two."

*   *   *

The remainder of Romans' life is one of America's mysteries. Seeing Bernard Romans totally disappear from view may frustrate the reader, who wants to know "how it all came out." That is how this biographer felt as he watched the possible sources of information on Romans' life trickle down— and then completely dry up. The "ingenious man," around whom had swirled some of the earliest, angriest and best documented quarrels of the American Revolution, abruptly is no more— neither in book, nor map, nor letter nor reference. Romans' noisy revel on the American historical stage is suddenly over, "melted into air, into thin air."

That ancient silence is broken only by a belated application from Elizabeth Whiting Romans for her

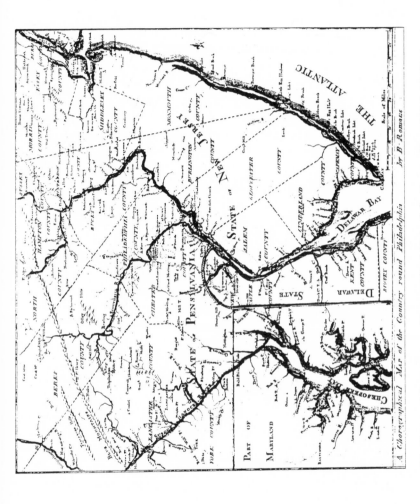

36. Romans' foot-square *Chorographical Map of the Country Round Philadelphia* 1778.

husband's "seven years half-pay," plus the widow's pension legally due to her. It was filed in her 87th year before Charles J. Dodge, a New York judge of the Court of Common Pleas, on October 15, 1846!

Mrs. Romans' affidavit, so long after the fact, sets forth her recollection of how in the summer of 1780, after two years' respite as a civilian in Wethersfield, her husband again volunteered, or was "ordered, to go to the State of South Carolina, there to join the Southern Army. Shortly thereafter, he sailed from New-Haven or New-London, for the place of his destination." [This may possibly have some connection with the fact that Horatio Gates had just been commissioned by Congress to lead a Southern Army against Cornwallis, using a core of Continentals detached from Washington's forces; within two months, Romans' former commander would suffer unmitigated disaster at Camden SC.]

"The vessel and crew with which he embarked were captured by the British," Mrs. Romans' affidavit continues [on May 12th Charleston, from whence Romans had sailed north seven years before, fell to a British siege], "and her husband was carried as a prisoner of war to Montego Bay, Island of Jamaica, where he was held in captivity until the close of the war in 1783. The British authorities were applied to deliver him up by exchange, which they refused to make, on account of said Romans' ability to do so much injury to the British interests."

If Elizabeth Romans' information was accurate, her husband had become one of perhaps eight thousand patriot seafarers captured on the high seas by the British during the war. But Royal Navy prisoners taken off the East Coast were usually sent to temporary Caribbean detention in Barbados, St. Lucia or the Leeward Islands— not Jamaica— before shipment to the terrible hulks in New York's Wallabout Bay, or somewhat milder incarceration at England's Mill and Forton prisons, near Plymouth and Portsmouth, where the order of the day was largely attempts at escape into the English countryside, and subsequent recapture.

Romans' renewed participation in the Revolution may have been in trade for his long-sought patriot colonelcy; if so, it saved him from the bitter fate of most rank-and-file American prisoners, but also helped hoist him on his own petard. At that time, colonels in the British army held an administrative, not a combat role, which seriously diminished the possibility of any one-for-one prisoner exchange (until late in the war when a colonel was finally determined to be worth six captains— or 100 privates or militiamen).

Unlike the army's records, accurate naval prisoner-of-war reports are practically non-existent; the Royal Navy could hardly have anticipated our intense 20th century demand for historical information. Parliament did not even legally define the Navy's captives as POW's until March 1782; prior to that they occupied some sort of extra-legal existence, to avoid any semblance of de facto recognition of the rebellious colonies.

Thus any possible exchange of a valuable naval prisoner who had been taken along the Atlantic coast by British warships and brought permanently to "Montego Bay, Island of Jamaica," was ringed by a complicated series of military and civilian considerations, practical and administrative difficulties and private ambitions. Even the most clearsighted and forceful naval commanders and West Indian governors were repeatedly overborne by such pressures.

"She further believes," Mrs. Romans' affidavit continues, "that her husband was shipped by the British authorities thence to some port in the United States, and he was said to have died on his passage, although from circumstances attending his demise, his friends had good reason to believe him to have been wilfully murdered."

Mrs. Romans' pay and pension claim was disallowed on indefensible grounds: that her husband's services were never military— even though her affidavit clearly states

how Romans "was also sent on a northern expedition" in
1778. A diligent search of existing army records in
Washington or Harrisburg would have disclosed that
fact.

Elizabeth Romans died two years after filing her claim,
having outlived her husband by more than 60 years, and
their son by 44. An obituary notice in the *New York Daily
Tribune* for May 13, 1848 "respectfully invites her relatives
and acquaintances, and her grandson B.H. Romans [the
family line continued], to attend her funeral at No. 25
Bayard Street."

The formal British biographical entry on Bernard
Romans offers a completely different, equally unsub-
stantiated story of the last half-decade of Romans' life. It
states "he was captured in 1779, probably at Stoney Point
on the Hudson," a post that changed hands three times
during six weeks of that year, "and was sent to England."
But no American was captured at Stony Point in any of
those exchanges; the British always occupied an abandoned
post. There is no record of Romans' going anywhere near
Stony Point (many miles from Wethersfield and not even
labeled on any of his Hudson River maps) during the
summer of 1779. His wife was six months pregnant, and he
had left the army.

The British biographical account continues: "His exchange
was refused, and after the peace he again practiced in
England as an engineer [a statement for which no British
record can now be found]. In 1784 he sailed for New York,
carrying with him a large sum of money, and, as he was
never heard of again, is supposed to have been murdered
during the passage." (Romans often carried large quantities
of money on his person; in 1778 he had advertised for $390
in Continental bills "wrap'd up in a letter" that he had lost
between Hartford and Wethersfield.)

There is a highly unlikely possibility, not previously
considered, that Romans defected to the enemy in the
summer or fall of 1780, under far less dramatic circumstances

but certainly at the same moment as his erstwhile patron and comrade-in-arms, General Arnold. If so, why (unlike Arnold) did Romans leave his young wife and infant behind?

On the other hand, if Romans was indeed a British naval prisoner in Jamaica for four long years, should he not have corresponded at least once with Elizabeth? The indefatigable letter writer of Constitution Island seems suddenly struck dumb. It is all rather perplexing. In what was probably his 64th year, energetic, adventurous, crusty Bernard Romans became one of America's earliest unknown soldiers.

37. Ruins of battery on Constitution Island, looking east.

38. "Romans Battery" plaque at Trophy Point, West Point Military Academy.

Epilogue

What eventually became of Fort Constitution, Bernard Romans' envisioned Gibraltar-on-the-Hudson? On October 7, 1777, a year and a half after Stirling's scathing report to Washington, and after countless alarums and excursions by the British along the river, a combined enemy force of four thousand men under Generals Henry Clinton, William Tryon and naval Captain James Wallace— ostensibly moving (unsuccessfully) to relieve beleaguered Burgoyne at Saratoga— landed near Stony Point, swept around Bear Mountain and down the shores of Popolopen Creek, and poured over the defenders of Forts [James] Clinton and Montgomery. They sent a flag north to what remained of Fort Constitution, demanding immediate surrender.

Captain Gershom Nott (also from Wethersfield), commanding the post's usual inadequate garrison, cavalierly fired on the British emissary. Once darkness fell, however, instead of mounting some legendary resistance *a la* Baltimore's Fort McHenry 37 years later (thus resolving for history some idea of the effectiveness of Romans' work), Nott demonstrated his true belief in the strength of the fortifications by setting fire to everything that could burn and abandoning the island— unhappily leaving all his 77 guns unspiked.

The next morning General Tryon, "to correct the insult to his flag of truce," sailed up the river from the ruined works above Popolopen Creek [three years earlier, Tryon had been one of the first subscribers to *A Concise Natural History of East and West Florida*]. The General's advance party

145

"found the Fort evacuated in the greatest confusion." His troops clambered ashore and with powder, axe and crowbar destroyed whatever remained of Romans' and others' later handiwork.

In less than two days of bloody fighting, British destruction of all the "Fortifications at the Highlands," the posts that had had cost the young country so much toil, argument, equipment and more than a quarter of a million dollars, was complete. The only reaction from Bernard Romans was a brief inset mention of Tryon on the 1778 *Map of the Northern Department,* calling the Governor-turned-general a "Prince of Land Jobbers."

A British naval detachment of six ships under Captain Wallace carried General John Vaughan and 1,600 soldiers 35 miles farther upriver to Esopus landing and Kingston, seat of New York State's revolutionary government. That area, according to General Vaughan, was "a nursery for almost every villain in the Colony."

Receiving scattered fire from a few of Kingston's inhabitants, the British burned the town to the ground on October 16, 1777, "laying everything in a rubbish of ashes." It was an empty gesture and of little use to Burgoyne; that unfortunate general was already preparing to capitulate at Saratoga. Hearing news of Burgoyne's surrender, General Vaughan sent word to Clinton on October 27th (from his somewhat inappropriately-named flagship *Friendship*) that it was "impracticable to give further assistance." Whereupon he re-embarked the British "relief" force at Esopus, picked up his temporary Highlands' garrisons (one of the forts had already been renamed "Fort Vaughan"), and sailed downriver to New York. After 20 days in enemy hands, West Point was again— and has since remained—free.

Despite the glorious victory to the north, the waste of Fort Constitution, plus new losses in the Highlands and elsewhere, preyed on Washington's mind. Voicing these thoughts, the outwardly imperturbable leader soon wrote from Valley Forge to his favorite Major General, the 20-

year-old Lafayette: "We must not, in so great a contest, expect to meet with nothing but sunshine."

There are many monuments on the grounds of the United States Military Academy to honor soldiers of the American Revolution. Only two words on a bronze plaque at Trophy Point, on the bluff overlooking Constitution Island and the narrow river gut now called "World's End," pay any tribute at all to the efforts of feisty Captain Bernard Romans, who— for whatever coupling of patriotism and ambition— came forward at a critical moment to help his adopted country.

The gods loved the young American heroes who led our earliest resistance to the Crown. Joseph Warren, Richard Montgomery and Nathan Hale were carried off in the glory of their apparent perfection. But even the greatest of those who survived revealed flaws during the grim years that followed; Romans was merely a striking example.

Yet despite his unusually generous self-interest, the cantankerous old captain was first and foremost a patriot. To seek some appropriate eulogy, we need look no further than the General Order addressed to all patriots by George Washington on April 18, 1783, exactly eight years after Lexington and Concord:

"While the General recollects the almost infinite variety of scenes through which we have passed, he cannot help wishing that all the brave men (of whatever condition they may be) who have shared in the toils and dangers of effecting this glorious revolution, might be impressed with a proper idea of the dignified part they have been called upon to act, on the stage of human affairs.

"For, happy, thrice happy, shall they be pronounced hereafter, who have contributed anything, who have performed the meanest office, in erecting this stupendous *fabric of freedom and empire* on the broad basis of independency; and who have assisted in protecting the rights of human nature, and in establishing an asylum for the poor and oppressed of all nations and religions."

39. MS *Fernpassat* (Monrovia) passing Romans' Battery ruins (1984).

Bibliographical Notes

Primary historical sources for this period of the American Revolution are well-known to scholars. Avocational historians (like myself) who may be interested in casting their own net will find the sources for quotations or letters noted either within the text or below.

Bernard Romans' *A Concise Natural History of East and West Florida*, first published by James Rivington in New York in 1775 (although without a publisher's imprint), was reset and reprinted almost two centuries later by the Pelican Publishing Co. [*New Orleans, 1961*] with an introduction by Louise Richardson. The following year the University of Florida Press [*Gainesville, 1962*] issued a facsimile edition of the original 440-page book, with a longer introduction by Rembert W. Patrick. Both reprints are available in major reference libraries.

Two centuries after its initial publication, Romans' discursive natural history of the Florida wilderness (hardly "Concise") still continues to provide informative reading "for the Sage in his Cabinet and the Mariner in his Ship." The book was issued as a single volume; the promised second volume never appeared and probably was never written.

Both volumes of Romans' two-part *Annals of the Troubles in the Netherlands, etc.* were published, but only a handful of copies of the second volume survive, making it one of the rarest of early American books. The New York Public Library and the Connecticut Historical Society each possess a set of both volumes bound as one.

Romans' huge 40-square-foot (1772) unpublished manuscript map with its "attempt at a short description of West Florida," has mysteriously disappeared following its transfer from London's Colonial Office (cataloged in 1910 as CO-700-52) to the collections of the Public Record Office. Fortunately, a photographic copy is preserved in the map collection of the Library of Congress. The earlier (c. 1770) manuscript map of the East Florida peninsula from the Keys to 30° N. [*1:750,000, Figure 1*] may actually be part of the loan "security" required from Romans by the New York Marine Society. If so, the map may have been appropriated from the Society by Lieut. Gen. Sir Henry Clinton for use in his southern military planning during the Revolution; it was found among that general's papers and is now at the University of Michigan's William L. Clements Library.

Almost all of Bernard Romans' published cartography, including British versions, is in the Library of Congress. A few originals and several reproductions are in the Map Room of the New York Public Library and

149

the collections of the Connecticut Historical Society. They include miniaturized and full-sized copies of the impressive small-scale [*1:600,000*] whole sheet maps that accompanied the *Concise Natural History* issued by Pelican in 1962. Charles Evans' authoritative *American Bibliography* [*Chicago, 1909*] notes that "the two maps missing from all copies of this work were afterwards [re]engraved by John Lodge, and published in London in May 1781; copies are sometimes found with these maps inserted."

Plans by the Florida State Historical Society in 1920 (the bicentennial year of Romans' supposed birth) to issue a reprint of the *Concise Natural History* never materialized. Instead the Society published, four years later, *Notes on the Life and Works of Bernard Romans* [*Deland FL, 1924*] in a limited edition of 325 copies. That study, accompanied by full-sized reproductions of Romans' East and West Florida maps, was prepared by P. Lee Phillips, Custodian of Maps for the Library of Congress. Phillips considered Romans a "universal genius"; his *Notes* contains a carefully annotated bibliography of all the Library's Romans holdings. It reprints, among other valuable items, the full text of Mrs. Romans' 1846 pension affidavit.

A handsome facsimile edition of the *Notes* (and whole sheet maps) with a detailed introduction by John D. Ware was published by the University Presses of Florida [*Gainesville, 1975*] to celebrate the American Revolutionary Bicentennial.

Romans' (and his publishers') advertisements in many New York and New England newspapers— notably New York's *Gazetteer,* and *Journal, or General Advertiser*; Boston's *Gazette, and Country Journal, Post-Boy and Advertiser,* and *Weekly News-Letter*; the *Massachusetts Gazette*; Connecticut's *Gazette, and the Weekly Intelligencer, Courant, and the Weekly Intelligencer, New-London Gazette,* and *Journal*— all offer (usually immodest) information about Romans' cartographic merchandising efforts from 1774 to 1778.

General Haldimand's quoted observations on the barrenness of the Floridas are contained in a January 1768 letter from Thomas Gage to Viscount Barrington at the War Office, part of General Gage's printed *Correspondence* [*New Haven, 1931*].

H.S. Halbert's critique of the *Map of Mississippi* produced by Romans after his 1772 exploration of the Tombigbee Basin, was delivered to the annual meeting of and reprinted by the Mississippi Historical Society [*Jackson, 1902*]. Romans' journal of that trip appears both in the *Concise Natural History* and Peter J. Hamilton's *Colonial Mobile* [*Cambridge MA, 1910*].

Dr. Alexander Garden's comments to James Ellis on Romans' lack of formal botanical training are from James Edward Smith's *Correspondence of Linnaeus and Other Naturalists* [*London, 1821*]. Romans' own correspondence with John Ellis is from the collections of the British Linnaean Society; it includes the engineer's reference to a trip to Cartagena (in Colombia). The editorial comment on Charleston professional life is from *The South Carolina Gazette* in 1775.

The minute books of the Standing Committee of the New York Marine Society rest in the Luce Library of the Maritime College of the State University of New York. That unique library is located deep within the converted casemates of Fort Schuyler, the superannuated massive granite structure (directly beneath the Throgs Neck Bridge over Long Island Sound) conceived by some latter-day Bernard Romans to guard New York City against any 19th century naval incursion.

Romans' earliest descriptions of Florida for the Colonial Office, as well as his 1773 letters to Lord Dartmouth and Dr. Hugh Williamson regarding the proposed trip around the world, are in London's Public Record Office and the collections of the British Historical Manuscripts Commission. The details of Romans' on-and-off "pension," his reward for explorations in West Florida, are covered in the Colonial Office Series *Documents of the American Revolution, 1770-1783* [*Dublin, 1974*].

Dr. Warren's eloquent cry for resistance against British oppression is from the Suffolk Resolves. Romans' autobiographical defense against the attack by John Scott was printed in the *New-York Gazetteer* of February 10, 1774, and contains additional detail on his years in the South; it is quoted herein at face value. De Brahm's suggestion to turn British North America into a garrison state is from a 1774 letter to Lord Dartmouth in the London collection of the Historical Manuscripts Commission.

Rev. Ezra Stiles' comment on Romans (and his "trip to Labradore") is from Volume I of Stiles' *Literary Diary* [*New York, 1901*]. Romans' letter to William Bradford regarding the use of good-quality paper for his maps is in the manuscript collection of the American Philosophical Society in Philadelphia. His August 1774 promissory note (£27 5s. 1d. for two hogsheads of rum) is in the New-York Historical Society.

A contemporary sketch of Abel Buell by John Warner Barber is in the *Connecticut Historical Collections* [*Hartford, 1836*]; that colorful and gifted engraver has also been the subject of more recent studies. Ebenezer Hazard's criticism of the *Concise Natural History* is from the Belknap-Hazard Letters in the *Massachusetts Historical Society Collections* [*Boston, 1792 etc.*].

For providing valuable shortcuts through widely scattered manuscript archives [or related microfilm], primary research scholars in American Revolutionary history owe an incalculable debt to Peter Force's (1790-1868) incomplete but nevertheless monumental *American Archives* [*Washington, 1837-1853*], one of the most ambitious schemes of historical collecting, editing and publishing ever undertaken. Force's oversized nine-volume, multi-million-word "Series 4 and 5"— culled from his collection of nearly 23,000 historical books, 40,000 pamphlets, 250 volumes of newspapers, 1,200 maps and views, 429 volumes of original manuscripts and 360 volumes of transcripts, gathered over 35 years (and purchased by the Library of Congress in 1867)— is available at most major libraries. Force's volumes reprint thousands of priceless documents from the earliest Revolutionary period.

The full work planned by Force, a newspaperman, avid manuscript collector and one-time mayor of Washington DC, was intended to cover all the extant written materials from the discovery of North America to 1789. It was sponsored by the federal government until funding was eliminated by William L. Marcy (may we long remember his name), U.S. Secretary of State under Presidents Pierce and Buchanan; *Series 1, 2 and 3* were outlined, but never begun.

Force's six-volume *Series 4: A Documentary History of the English Colonies in North America from March 7, 1774 to July 4, 1776*— particularly Volume II— contains many references to Bernard Romans' wartime services and much of his correspondence with the New York Committee of Safety, the New York Provincial Congress/Convention and the Continental Congress (both of whose manuscript *Journals* are available separately in state and national archives). Force also prints pertinent committee and military reports to and from those bodies. *Series 4, Volume II* also covers the flareup between the Fortifications Commissioners and the Hudson River skippers.

Force's [incomplete] three-volume *Series 5*, commencing with the Declaration of Independence (and originally planned to end with *The Definitive Treaty of Peace with Great Britain, September 3, 1783*) never moves beyond December 31, 1776— thus tantalizing generations of scholars seeking detailed documentation for the rest of the American Revolution in one convenient place. *Series 5* also contains a few references to Romans' service in the Canadian and Northern Departments.

The warning against the unfortunate outcome of burgeoning patriot violence is by conservative Joseph Galloway, as he renounced his appointment to represent Maryland at the Second Continental Congress. General Thomas Gage's letter to Captain William Delaplace is in the Gage Papers at the Clements Library. The solemn observation on the "surprize of Ticonderoga" is from Dodsley's *Annual Register, or a View of the History, Politics and Literature for the Year 1775* [*London, 1777*], an incredibly astute and continuing evaluation of military aspects of the Revolution. Established by Edmund Burke in 1758, the *Register*— heavily plagiarized by at least nine 19th century American historians— reflects considerable patriot bias.

The balance of the Ticonderoga/Fort George story is drawn from Captain Edward Mott's *Journal* [in Volume I of the Connecticut Historical

Society Collections] and Benjamin Franklin De Costa's *A Narrative of Events at Lake George* [*New York,1868*]; *Notes on the History of Fort George, etc.* [*New York, 1871*]. Benedict Arnold's comment on Ethan Allen is from Force *Series 4, Volume II.* The MS. accounting of Romans' expenses on the Ticonderoga-Fort George expedition is in *Series I, Volume III* of the Connecticut State Library Archives. Captain John Nordberg's December 1775 petition to the Provincial Congress is in the Office of the New York Secretary of State.

Rev. Timothy Dwight's characterization of the Hudson Highlands is from Volume III of his *Travels in New-England and New-York* [*New Haven, 1822*]. Washington's comment on early military engineers comes from *The Writings of George Washington From the Original Manuscript Sources, 1745-1799* [*Washington DC,1931-44*]. Charles Carroll of Carrollton's observation on the same subject is in Force's *Series 5.* Colonel Putnam gives Crailac as his inspiration for the chandeliers on Dorchester Heights in *Memoirs of Rufus Putnam and Certain Official Papers and Correspondence* [*Boston, 1903*].

Capt. Edward C. Boynton's handsome *History of West Point and Its Military Importance* [*New York, 1864*] relies completely on Force's *Archives* to tell the Fort Constitution story. Parts of Chapter 2 of Lt. Col. Dave Richard Palmer's excellent recent history of Fortress West Point, *The River and the Rock* [*New York, 1969*] cover Bernard Romans' stormy involvement with the Fortifications Commissioners. Lt. Col. Merle G. Sheffield's *The Fort That Never Was* [*West Point, 1969*] touches the same ground, and offers an interesting analysis of planned but uncompleted military structures on Constitution Island.

John Berrien's comment on General Wooster is from a letter in the New-York Historical Society. Notes on the Continental Congress debates over Romans' plans come from John Adams' and Richard Smith's *Diaries,* in the Massachusetts Historical Society and the Library of Congress, respectively. The report to John Hancock from the Congressional Committee to the Northward is in the National Archives. "Moore's Folly" is so labeled on the river vignette of Thomas Jeffreys' engraving *A Prospective View of the BATTLE fought near Lake George, etc.* [*Boston, 1755*]. Jefferson's observation is drawn from *Volume I* of his papers [*Princeton,1950*].

British naval intelligence references to Bernard Romans and Fort Constitution are in Admiralty papers reprinted by our contemporary version of Force's *American Archives*— the equally ambitious and massive Department of the Navy, Naval History Division's, *Naval Documents of the American Revolution* [*Washington, 1966 etc.,* currently eight volumes through 1777 alone!]. The minutes of the February 1776 Board of Trade meeting are from Volume 14 of the *Journals of the Commissioners for Trade and Plantations* in London's Public Record Office, which also holds Lord Germain's critical August 1776 letter to Governor Chester.

Robert R. Livingston's letter to Thomas Lynch is in the New York State Library. The story of the April 1776 visit to Fort Constitution is drawn both from the *Journal of Charles Carroll of Carrollton* [*Baltimore, 1876*] and Volume 22 of Benjamin Franklin's *Papers* [*New Haven, 1982*] (Franklin's earlier comment on Lord Dartmouth is also from those *Papers*). General Horatio Gates' observations are from the Gates Papers in the New York Public Library. The remark about Gates' vanity was made by Robert Troup, in *Sibley's Harvard Graduates* [*Boston, 1873*]. The reference to the destruction of Kingston comes from a Ms. diary kept by Abraham Hasbrouck.

Romans' and Hancock's manuscript letters to General Schuyler are in the Lossing section of the Emmet Collection in the New York Public Library. The 1779 Glastonbury letter (tipped into a 1924 Goodspeed's [Boston] auction copy of the *Concise Natural History*— present whereabouts unknown) is available on New York Public Library Prints Division (*Archives of American Art*) microfilm.

The perplexing reference to Romans' continuing his practice of engineering in England 1779-1784, presumably as an American prisoner of war, is from a entry prepared during the 19th century by Gerald le Gruys Norgate for *Volume XVII* of George Smith's [British] *Dictionary of National Biography,* a standard reference work in most American libraries. Despite the lack of any substantiation for this Norgate reference anywhere in Great Britain, it soon became popular early 20th century historical hearsay in America.

For example, that *Dictionary* reference is transcribed word-for-word in Samuel Abbott Green's *Ten Fac-simile Reproductions* [*Boston, 1903*], which also reproduces Romans' rare *An Exact View of the Late Battle at Charlestown.* Four years later David McNeely Stauffer covered the work of Abel Buell, Robert Aitken and Romans in his *American Engravers* [*New York, 1907*]— and copied Green's same information. An aesthetic and technical appreciation of Romans' *Exact View* is in *A Descriptive Catalog of an Exhibition of Early Engraving in America,* published by the Boston Museum of Fine Arts in 1904.

Washington's Valley Forge comment to Lafayette (page 147) is from Jared Sparks' 12-volume *The Writings of George Washington* [*Boston, 1834-7*]. All Americans would all do well to ponder it from time to time.

Benson John Lossing's *Pictorial Field Book of the American Revolution* [*New York, 1851*] mentions an heirloom miniature of Mrs. Romans, "beautifully" painted by her husband, but its present whereabouts are unknown. One of the ironies of American history is that no likeness of the engineer-cartographer-artist-engraver himself is known to exist in any collection in the United States or Great Britain.

Unless one cares to consider the mysterious half-hidden horizontal profile in *Figure 26a* as a strange Bernard Romans self-portrait.

Index

157

158